CAMBRIDGE LIBRARY COLLECTION

Books of enduring scholarly value

Physical Sciences

From ancient times, humans have tried to understand the workings of the world around them. The roots of modern physical science go back to the very earliest mechanical devices such as levers and rollers, the mixing of paints and dyes, and the importance of the heavenly bodies in early religious observance and navigation. The physical sciences as we know them today began to emerge as independent academic subjects during the early modern period, in the work of Newton and other 'natural philosophers', and numerous sub-disciplines developed during the centuries that followed. This part of the Cambridge Library Collection is devoted to landmark publications in this area which will be of interest to historians of science concerned with individual scientists, particular discoveries, and advances in scientific method, or with the establishment and development of scientific institutions around the world.

Memorials

These *Memorials* of Andrew Crosse (1784–1855), published by his wife after his death, include his experiments, and some of his poetry and prose. After graduating from Brasenose College, Oxford, in 1805 (described in this volume as 'a perfect hell on earth'), he returned to his family's manor house where he studied electricity, chemistry, and mineralogy, and installed a mile and a quarter of insulated copper wire in his grounds. A controversial figure, Crosse was thorough in his approach to his scientific work, if somewhat unusual in his practice. In 1836 he famously conducted a series of experiments on electro-crystallization in which he noted an appearance of life forms, named *Acarus*, seemingly created in the metallic solutions which should have been destructive to organic life. This book recounts these experiments, and the public sensation that they gave rise to by their apparent suggestion of life created by electricity.

Cambridge University Press has long been a pioneer in the reissuing of out-of-print titles from its own backlist, producing digital reprints of books that are still sought after by scholars and students but could not be reprinted economically using traditional technology. The Cambridge Library Collection extends this activity to a wider range of books which are still of importance to researchers and professionals, either for the source material they contain, or as landmarks in the history of their academic discipline.

Drawing from the world-renowned collections in the Cambridge University Library, and guided by the advice of experts in each subject area, Cambridge University Press is using state-of-the-art scanning machines in its own Printing House to capture the content of each book selected for inclusion. The files are processed to give a consistently clear, crisp image, and the books finished to the high quality standard for which the Press is recognised around the world. The latest print-on-demand technology ensures that the books will remain available indefinitely, and that orders for single or multiple copies can quickly be supplied.

The Cambridge Library Collection will bring back to life books of enduring scholarly value (including out-of-copyright works originally issued by other publishers) across a wide range of disciplines in the humanities and social sciences and in science and technology.

Memorials

Scientific and Literary, of Andrew Crosse, the Electrician

CORNELIA CROSSE

CAMBRIDGE
UNIVERSITY PRESS

CAMBRIDGE UNIVERSITY PRESS

Cambridge, New York, Melbourne, Madrid, Cape Town, Singapore,
São Paolo, Delhi, Dubai, Tokyo, Mexico City

Published in the United States of America by Cambridge University Press, New York

www.cambridge.org
Information on this title: www.cambridge.org/9781108014915

This edition first published 1857
This digitally printed version 2010

ISBN 978-1-108-01491-5 Paperback

MEMORIALS

OF

ANDREW CROSSE.

MEMORIALS,

SCIENTIFIC AND LITERARY,

OF

ANDREW CROSSE,

THE ELECTRICIAN.

LONDON:
LONGMAN, BROWN, GREEN, LONGMANS, & ROBERTS.
1857.

Dedication.

TO

SIR RODERICK IMPEY M'URCHISON,

D.C.L., F.R.S., ETC. ETC.

DIRECTOR-GENERAL OF THE GEOLOGICAL SURVEY OF GREAT BRITAIN.

DEAR SIR RODERICK,

I SOUGHT, and you kindly permitted me, the privilege of dedicating the following pages to you, on the ground of the interest which you take in Science generally.

Though you disallowed that you possessed any especial knowledge of Electricity, yet permit me to say that you have ever been foremost among the philosophers of this epoch in receiving a class of experiments whose relation to Geology is rather a matter of prophecy than of acknowledged fact.

Your kindly feelings of personal friendship for my late husband, and your appreciation of his character, will, I am sure, lead you to understand the anxious desire I have to preserve even this brief and imperfect record of Andrew Crosse.

On my own part I beg you to accept this slight tribute, as a mark of the admiration I feel for one who has done so much in advancing the new-born science of the century,—in determining, not only the Geological nature of the Earth's strata, but the Geographical character of its surface.

With these sentiments of respect, allow me to subscribe myself,

Yours very sincerely,

CORNELIA A. H. CROSSE.

PREFACE.

THE following pages only assume to be Memorials linked together. For a Biography, or even Memoir, there were neither materials nor occasion. The endeavour has been to give such matter alone as may be interesting for the public to know of Andrew Crosse, than whom few so deserving were so unknown. His fame would have been wider, had he been poor, ambitious, or wise in worldly wisdom. He was none of these things.

For myself I claim some indulgence at the hands of criticism, inasmuch as memory and affection, rather than confidence in my own powers, have supported me through the prosecution of the task.

The details of my husband's experiments have cost me no small trouble to collect; for his notes were made, most generally, on loose scraps of paper, and to arrange them required an intimate knowledge of the trains of thought which were in his mind when he performed many apparently isolated or dissimilar experiments. It has not been in my power to consult any scientific friend of his early days; and I know that much has been lost that would now have been to me most valuable.

It must, however, be observed that every particular which is stated in these pages, connected with scientific discovery, can be authenticated by Mr. Crosse's own hand-writing. It is for the manner, the order, and the omissions in the arrangement that I claim indulgence.

It has been my endeavour not to intrude any personal feelings in this narrative. My husband's friends know how he was beloved by those near and dear to him. The public will, I trust, forgive me, if unconsciously I have fallen into the error I wished to avoid.

A small portion only of the posthumous poems of Andrew Crosse are now published; their reception must be left to the taste of the day, with this observation only, that, had their Author been living, he would most probably have revised and amended these poems before publication; whereas I do not feel either justified or equal to do what now, alas! must be left undone!

With apologies for intruding myself upon the reader's notice, there remains to me only the duty of authenticating these pages by subscribing my name.

CORNELIA A. H. CROSSE.

Comeytrowe House, near Taunton:
March, 1857.

CONTENTS.

MEMOIRS

OF

ANDREW CROSSE.

CHAPTER I.

PARENTAGE AND EARLY LIFE.

1784—1805.

About the year 1816, a party of country gentlemen were dining at Alfoxton Park, in the west of Somersetshire. The squirearchy of this period were wont to indulge in endless discussions about game and guns, horses and dogs, poachers, and the business of the petty sessions.

The convulsions of Europe had subsided, leaving people generally in that happy state of security when local annoyances and country news have claims to an overwhelming share of interest. But at this hospitable board there were other matters canvassed. The intellectual impulse which renewed intercourse with foreign states ever gives to our national and insular mind was making itself felt even in remote neighbourhoods. Some chance expression, resulting perhaps from the circumstance of this feeling, roused the hitherto most silent person of the party — a shy, but intellectual looking man, who appeared even younger than he was; a vehement utterance struggled in him with an embarrassment of manner at once singular and interesting. The allusions of his companion to some of the scientific discoveries of the day touched on his own peculiar train of thought; at least so it seemed, for he started up from his abstraction, and spoke eloquently of the advances which science had made. Forgetting his shyness and himself, and rising into enthusiasm, he proceeded to describe the power of electricity and the range of its influence; and at length their startled attention was fixed by his pronouncing these remarkable words: — "I prophesy that, by means of the electric agency, we shall be enabled to communicate our thoughts *instantaneously* with the uttermost ends of the earth."

This announcement was received as a wild chimera, and would hardly have gained the consideration it did, had not the speaker convinced his audience of the fullness at least of his own conviction in the results of a speculation at once bold and unprecedented. But the palpable improbability — nay, apparent physical impossibility — of such an application of a mysterious, almost unknown agent, called forth more astonishment than comprehension, more ridicule than credence. Yet, absurd as the idea was then deemed, most of the party assembled have lived to see the fulfilment of those prophetic words, uttered forty years ago.

The individual who thus foretold the electric telegraph, is the subject of the following memoir, — Andrew Crosse, then unknown in the scientific world, living in great retirement and intellectual isolation.

At a time when electricity was little more than a name, he foresaw the importance of the science in its practical application, and recognised in its laws an explanation of physical phenomena. The part which he took himself in advancing his favourite science, his successful imitations of nature in the laboratory, and the theories he advanced, I will not now stay to estimate; but first rather pause to consider the antecedents which belong to him personally.

In early life, time, place, and circumstance contribute to mould the future destiny. Imagination is the great architect of character, and to the impressions of infancy can be traced the tendencies of after life. A birthplace is more than a name; it environs our childhood, and often serves to develope a poetic temperament, and a love of the beauties of nature, which otherwise might have lain dormant. Eminently calculated to create such influences was the birthplace of Andrew Crosse.

In the west of Somersetshire is a range called the Quantock Hills; it is a wild and picturesque locality, where red deer and black game find a safe retreat; where purple heath and golden furze make the autumnal scene glorious with regal colouring.

It is a district where you may walk for miles without passing even a cottage, and where the whortleberry is the only crop. Yet these wild hills separate the two most fertile vales in England; nor are the hills themselves altogether barren, for in some spots the luxuriance of the vegetation and the gigantic growth of timber is not excelled in any part of the island.

The highest point of the Quantocks is upwards of 1400 feet above the level of the sea. At every step a varying prospect is presented to the view; and the denizen of the hills, breathing the pure exhilarating air, feels the world below him, and finds himself in the enjoyment of a free, wild solitude.

At the eastern side of the Quantock Chain is the parish of Broomfield, famed in Doomsday Book for the longevity of its inhabitants, and noticed in the history of Somersetshire for the magnitude and beauty of its trees. The old parish church is most picturesquely situated, and is in itself a perfect specimen of architecture suited to the scene. The time-worn walls chronicle the passing away of many centuries, and, gathered round in the sacred equality of death, the old tombstones mark the resting-place of many who were born and died in their parish. Some three or four cottages surround the village green, and close by stands the Manor House, the family seat of the Crosses, called Fyne Court, so named from the manorial fines having been collected there in olden days. It is an Elizabethan structure of moderate size, and was built in 1629 by one Andrew Crosse; but it has been so enlarged and altered by succeeding possessors, that it can boast only of its antiquity and irregularity. The undulating grounds are tossed together as if by some freak of nature, and the house itself is shadowed by lofty trees, through whose branches the winds never cease their sad music. The gigantic limbs and gnarled roots of the old beeches in the avenue strike one with admiration and with melancholy; the still vigorous trees, the growth of centuries, seem to mock at the short life of man; succeeding generations walk beneath the same shadow, and the narrow vault near by receives them successively, and their place knows them no more but as name-links in the old pedigree. Yet who is there can live within the walls where his forefathers have dwelt without feeling in some sort more than commonly united with the past? Sentiments of reverence impress us; the spirits of those who have gone before may still seem to hover about the old familiar place, and we cannot be surprised if superstitions attach themselves to the scene; — and truly at Fyne Court

> "A sense of mystery the spirit daunted,
> And said as plain as whisper in the ear —
> 'The place is haunted!'"

This was the birth-place of Andrew Crosse, the Electrician. But a few words about the family, who, like their house, were old and curious.

There is a tradition that the first of the family who settled in England was a Norman thane, who came over, of course, with William the Conqueror, and whose name was Odo de Santa Croce.

Camden writes of one "William Crosse, Esq., de

Charlenge," who, at the time of Edward I., fought in the van of the English army against the Scots at the battle of Falkirk, A. D. 1298; also a Sir William Crosse of Charlenge, at the time of Henry V., was slain, whilst fighting under that monarch at the battle of Agincourt, A. D. 1415. The Crosses had for generations been settled at Charlinch previously to their purchasing the moiety of the Manor of Broomfield, in the same county, in 1630, and the above-mentioned were ancestors of the family. These records were gathered from memorials in the possession of the Cotgreaves of Netherligh, Cheshire.

An interesting document exists in the Crosse family, of a grant of arms to Sir Robert Crosse of Charlenge) the direct ancestor of the subject of this memoir), who was knighted on the field of battle. They are specified therein as a very ancient family, long bearing arms. The date of this document is 1601. It is signed by Camden the historian, then Clarenceux king-at-arms.* This same Sir Robert, it is believed, accompanied Sir Walter Raleigh in his explorations.

* The Crosse Arms are quarterly, arg. and gu. in the first quarter, a cross crosslet of the second. Crest:—A cross patée fitchée gu., between two wings arg., each charged with a cross crosslet of the first. Motto:—*Se inserit astris.*

The following is a monumental epitaph of one of these ancestors: —

M. S.
FRANCISCI CROSSE
EX EQUESTRI FAMILIÂ IN AGRO SOMERSETTENSE ORIUNDI,
COLLEGII WADHAMENSIS APUD OXONIENSES SOCII,
ET MEDICINÆ DOCTORIS;
QUI POST EXTERNAS REGIONES PERAGRATAS HÂC IN URBE
PRÆCLARÂ PER ANNOS DECEM PLUS MINUS MEDICINAM
FELICITER FECIT, TANDEM FEBRE CORREPTUS OBIIT
TERTIO DIE OCTOBRIS AN: DOM: MDCLXXV.
ET HEIC SEPULTUS JACET SUB
SPE VITÆ ÆTERNÆ.
VIR DOCTRINÂ ET ANTIQUÂ MORUM SIMPLICITATE CLARUS.

The pedigree of the Crosses makes mention of a marriage with the ancient family of Bottelor, Earls of Ormonde, and at a later period with a granddaughter of Lord Say and Sele (who was the grandmother of Andrew Crosse), a lady of exceeding worth, if we may credit the affectionate remembrances of her son.

In the life of Bampfylde Moore Carew, it is mentioned that he was hospitably entertained at Broomfield by Andrew Crosse, Esq. (who was the great uncle of the electrician). He was chairman of the Quarter Sessions—and a man of consideration in his time,—remembered for the soundness of his judgments, and somewhat celebrated for his skill as a musician. At his death in 1769, Richard Crosse, his nephew, the only son of the late rector of Cannington, succeeded to the estates. The new owner of Broomfield, the father of the subject of this memoir, had travelled much on the Continent at a period when the intercourse between nations was limited to the few; he spoke most of the European languages, and was altogether a most accomplished scholar. The old library at Fyne Court bears witness of his extensive knowledge of French and Italian literature. He frequented the Court of Lewis XVI. in the days of its extravagant splendour, when were being sown broadcast the seeds which yielded a strange harvest of barricades and guillotines. The Englishman was destined to see curious mutations; the brilliant scene shifted, court and courtiers were dispersed, the old régime was dying a violent death; a few years and Mr. Crosse mounted the tricolour, and stood upon the ruins of the Bastile the day it was captured.

He had long held liberal opinions and professed himself a republican in principle, though he had been high sheriff of Somerset only two years before these events. His views of the political changes of the day rendered him excessively unpopular; and, on one occasion when he returned from France, he was obliged to avoid the town of Bridgewater in his way home, for the populace stigmatised him with the name of Jacobin, and threatened to smash his carriage. This was a time when society was stirred to its very lees. We cannot marvel that England, fearful for the sanctity of her hearths and altars, redoubled the vigilance and severity of her laws, and looked with jealousy upon aught that savoured of liberal tendency. All violent antagonism of principles gives birth to a new order of things; the good comes not to the age which suffers, but is developed in the maturity of the future. "Every great man," says Hayzlett, "is the son of his age, but not its pupil." The examples of Toryism and Sans-culottism have produced constitutional reformers, such as Andrew Crosse became in after life. The *time* of our birth is never without its influence.

But to retrace our steps in the narrative: some years before the meeting of the States-general, before demolishing of the Bastile, before guillotining of king, queen, and royal family, Mr. Richard Crosse married, secondly, Susannah, daughter of Jasper Porter, Esq. of Blaxhold, in the county of Somerset, and on the 17th of June, in 1784, Andrew their first child was born. Mr. and Mrs. Crosse had one other son, born in 1786, and Mr. Crosse had an elder daughter, the offspring of his first marriage.

I have been told a singular instance of little Andrew's memory: he perfectly well recollected a dog called Rover; he used to describe where the dog had had his dinner, and many particulars about him. Rover died before he was two years old! Andrew had a profound respect, mingled with some fear, of his father, who appears to have been a strict disciplinarian, but never a harsh parent. Mr. Crosse's character might well call forth the respect of his son, for he had the reputation of the most unflinching integrity. So proverbial was this characteristic that persons several times remarked to the boy, "I like your father, he is such an honest man." The little fellow grew quite irritated at hearing this so often, and turning round on one occasion, said, "Sir, would you have me the son of a rogue?" "Young gentleman," replied the elder, "when you are grown up, you will know what I mean."

When about four years old, the little Andrew went with his parents to France. The following are extracts from the diary of his mother: —

Wednesday, Oct. 1st, 1788. — Left Broomfield with Mr. Brown, Mr. Crosse, my little Andrew, and two servants; drank tea, supped and slept at Piper's Inn. Andrew quite delighted with his journey. Thursday we left Piper's Inn, the two gentlemen in one chaise, Andrew, myself, and Sarah in the other; we called on Dr. Lovell at Wells, took fresh horses there, one of which occasioned an alarm, but we leapt out of the chaise, and insisted on taking another. We arrived comfortably at Bath, took a cold dinner at the Bear. Mem: We were charged four shillings for about three pounds of a breast of veal, and proportionally dear for other articles. We pursued our journey to Chipenham and slept there; the accommodations were good, but miserable attendants. Left Chipenham Friday morning, and came to Marlborough. * * We only stayed at Marlborough to take fresh horses, and went on to Newberry, where we took a cold dinner; and drank tea and supped at Reading. * * At Reading nothing appeared worth remarking, except the extravagance of the house and the affectation of the domestics. From Reading, Saturday morning, we set off with good weather and good spirits; stopped at Hounslow for fresh horses, and arrived in London about three o'clock. What a beautiful entrance through Hyde Park; the innumerable grand carriages, passing and repassing, together with such an immense crowd of people formed a scene too pleasing to be forgot. * * We stayed in London, Sunday and Monday. * * Tuesday morning we left and proceeded to Dartford, and from thence to Rochester, and where we took dinner, the ships in full sail on the river Medway was a delightful sight, and made my little Andrew ready to leap out of the chaise, so great was his joy. At Dover, we were detained three days by contrary winds. * * We left the Royale Hotel at Dover, Saturday morning, being called at five o'clock; we took a hasty breakfast, and went on shipboard about seven o'clock, intending to sail to Calais; but the wind blew quite contrary, the sea was very rough, and there was scarce a soul on board but was sick. After being near six hours on the sea, sailing against the wind, the captain proposed to the passengers to go to Boulogne, to which some of them reluctantly consented, and within two hours we were in sight of it. A boat was now to convey us on shore, and it was then I really began to tremble. My poor sick child was carried by French sailors into the boat, and I had no other way of making them know the value of him, but by the agony in which they saw me. The sail was up, and the boat much inclined on one side. But great was my joy when we saw forty or fifty women skipping and dancing in the water, with their petticoats above their knees; they presented themselves by the side of our boat to take us on their shoulders, and carry us on shore. Some of the women were of a gigantic make, and make nothing of carrying Mr. Brown and Mr. Crosse on their back; and Mr. Crosse very humorously knighted his lady by bastinading her with his sword. * * My little Andrew was joyous beyond expression, and having slept some hours aboard ship, played about the whole evening with uncommon alacrity. * * Sunday we attended the cathedral service at nine o'clock in the morning, and for the first time I saw high mass performed. * * We left Boulogne Tuesday morning, Oct. 14th, dined at Montreuil and slept at Bernai. Rose at six o'clock after almost a sleepless night, as poor Andrew was extremely ill. We breakfasted on milk from earthen porringers; and pursued our journey. * * We arrived at Amiens about six o'clock in the evening, at the Hotel Royale, which is a most excellent house. * * I congratulated myself on leaving this place; but I was soon taught how little cause I had to rejoice. By agreement between Mr. Crosse and the coachman we had quitted the public road to Orleans and travelled through by-roads, crooked and cross; and at last, towards evening, we came to an immense common, where the road was so uneven that we were every minute in danger of oversetting. Night came on, the coachman was ignorant of the way, my poor Andrew lay asleep on Sarah's lap, and to increase my anxiety the coach door was open, which I was obliged to hold in my hand, not without dreadful apprehensions that the dear child would be dashed out of the carriage with every jerk that it made. My silent prayers to the Almighty were heard for our protection, and we arrived at Chartreuse without any bodily accident. * * We left this place the next day in a conveyance something resembling a covered cart. In this manner ——." Here the MS. ends. The narrative of a journey performed not much more than half a century ago is a striking contrast to these days of railroads and electric telegraphs. The three score years and ten — the life of the "little Andrew" — has embraced a period of great change.

It appears that the family remained for some time at Orleans, for Andrew went to school there, and many are the childish anecdotes he remembered of the time and place. How he wandered away from his Swiss servant Barthelemy, and was found walking on the quay, lost in admiration of the ships; and was accidentally met by a French abbé, a friend of his father's. He could then speak French perfectly, but he entirely lost it afterwards, and was never a good modern linguist.

He returned from France with his mother, and from the age of six till eight he was with the Rev. Mr. White, who then resided at Dorchester. He was one evening walking out with this gentleman; they both stood gazing at a brilliant sunset; the child seemed much struck by the glorious flood of light which gilded the passing clouds, and Mr. White asked him what he was thinking about. "I was thinking," replied the little boy, "that this must be like the kingdom of heaven opening to all believers."

The little Andrew was scarcely eight years old when he left his kind friend and instructor Mr. White, but he had benefited so much by his teaching that he could read Greek tolerably well, and wrote a good Greek hand; he had not learnt Latin, and he wrote Greek before English. This circumstance affected his handwriting ever afterwards, his pen always flew very rapidly over a sheet of paper, which was the more singular as he wrote almost every letter detached. It was a great advantage to him to have had the benefit of Mr. White's tuition, for he was a finished scholar and a man of very elegant mind; he was the first person who ever signed a petition for the abolition of slavery, a fact which, I believe, is mentioned upon his tombstone at Huntshill, a parish of which Mr. White was rector for many years.

In February, 1793, Andrew went to school at the Rev. Mr. Seyer's, the Fort, Bristol. He surprised his master by reading Xenophon with tolerable facility. I have often heard him describe the effect of a large school upon his young uninitiated mind. Soon after he got there, a bigger boy tried to persuade him to tell a falsehood. "How I hated that boy," he observed, when recalling the incident; "I never was happy till I was strong enough to thrash him."

The following extracts from his mother's pocket-book tell their own tale. "February, 1st, 1796. My dear Andrew went to school, more dear and better beloved than ever. Never shall I forget his expressive looks at parting. May the God of Mercy bless and protect him! Thursday, parted with my beloved Richard, whose engaging sweetness of disposition endears him to every one who knows him. Happy prospect to look forward to these promising branches as a recompense for all my other troubles. Sweet children, both alike sharers of my maternal tenderness, my constant anxiety, and my daily prayers for their preservation and happiness! How delightful to be so beloved by children on whom the happiness of my life depends." A few pages further on, amidst the simple details of domestic life, are extracts from the writings of Madame de Genlis, Addison, Secker, Saurin, Pascal, Tillotson, Bruyère, Pope, &c. These indications of the character of her mind and the extent of her varied readings, prove Mrs. Crosse to have been a person of superior intellect. Her note books are a curious mixture of scraps from classical authors, receipts for puddings, bills of the sale of sheep, accounts of the number of pounds of butter used per week, and memoranda for sending for worsted for mending stockings! Every high, noble, and generous sentiment seemed to have found an echo in her heart; she used to say to her son, "never allow a mean or unworthy thought to enter your breast." Such was Andrew's mother; a true lady of the eighteenth century.

Andrew was a small, thin, but wiry boy, quick at his lessons, with a wild joyous temperament which delighted in fun and frolic. He was a great favourite with his schoolfellows, and his schoolboy friendships continued to the day of his death. Mr. Seyer's school has been rather celebrated for the names of those who were educated there. The late Rev.

John Eagles, who was for many years a contributor to "Blackwood," Mr. Broderip, the naturalist, John Kenyon, the friend of Wordsworth and Southey, the Hon. Henry Addington, the late Dr. Jenkyns, Master of Balliol and Dean of Wells, and several others who have distinguished themselves, were at school with Andrew Crosse.

Happy as Andrew was with his companions, yet never was a schoolboy more delighted to return home for the holidays. Even in advanced life his lively descriptions of the exuberant joy he felt on these occasions made one realise the picture of the greeting he gave his mother, as, tumbling head over heels into her presence, he seemed perfectly wild from excess of animal spirits; he loved his mother almost to idolatry; with the discernment common to childhood he appreciated her pure and noble sentiments, and felt the influence of her moral nature; to her, also, he was indebted for his first introduction to English literature; he very early began to devour books of travels, and to delight in whatever was strange and marvellous. His younger brother, Richard, was his playmate during the holidays; he had been sent at an early age to Dr. Valpy's at Reading. The two children were both very imaginative, and I have often heard Mr. Crosse say that they lived quite in a world of their own. They created a new language, and peopled their world with creatures like large fir cones, whom they called hoblegees; they used to imagine themselves pursued through the long passages of Fyne Court by these beings, whose mode of locomotion was something quite eccentric. They were not objects of horror. I have often been amused at the details I heard of their "Model Republic." Truly

"Heaven lies about us in our infancy."

Mingled with this childish nonsense were glimpses of profound wisdom. They made peculiar laws, language, and institutions, and the whole idea was pursued with an earnestness which made these fancies seem almost more truthful than actual realities. This waking-dream world continued to exist for several years in the brains of the two boys; but having no Gibbon to record its decline and fall, like greater things, it passed away.

Never was a boy more full of fun than Andrew. Some of his school anecdotes illustrate Wordsworth's expression which he himself thought so true, namely:

"The boy is the father of the man."

"I remember," he said, "a schoolfellow came begging me to give him a translation of '*M*[illegible] *tu-*

tissimus ibis' (the middle course is the safest). I gave him, 'The stork is safest in the middle of the pond;' Seyer did not appreciate the pun, and of course the boy got a caning. I myself was caned, upon an average, three times a day for seven years, but never once flogged. I had an irrepressible trick of laughing, which Seyer could never forgive. He was an admirable classic, a good grammarian; he had some nobility of feeling, was perfectly honest, but was a narrow-minded man, and without any sense of justice. I remember one day I was had up as usual to read Virgil: I had nearly completed the fifth book when I made a mistake in a word. 'Let me look,' said Seyer, and taking the volume from me he found that the whole of the fifth book had been torn out. I had repeated it from memory. I then explained to him that, without any fault of my own, one of my schoolfellows, in a fit of mischief, had torn it out some months since, and I confessed I had not had it during the half-year. My master's only reply was a good caning; and what was worse, whenever he was out of temper with me, he would call me up, and asking to look at my Virgil, repeating the caning every time."

There was a trick of his schoolboy days of which his old friends used often to remind him. I have often heard Mr. Kenyon laughingly tell him that in all his experiments he never did anything better. I give the story in Mr. Crosse's own words:—"I was always very fond of making fireworks. One day, while learning my Virgil, I continued to carry on the business of pounding some rocket mixture; but, as ill luck would have it, Seyer discovered my twofold employment, and immediately took away the mixture from me in considerable wrath. I watched where he put it; it was on the window-sill of a room which was always kept locked; the window, though not glazed, had close iron bars through which nothing could pass: the case was hopeless; I could not recover my rocket mixture, but a happy thought struck me, I was resolved that no one else should enjoy the spoil which I regarded as so valuable. I had a burning-glass in my pocket, and I thought of Archimedes and the Roman fleet; the sun was shining, and I soon drew a focus on the gunpowder, which immediately blew up. It was well that the house was not set on fire: as for me, I was reckless of all consequences."

"During the ten years that I was at school," said Andrew, "I never had enough to eat. I hated my schoolmistress, at whose table I unluckily sat, for she half starved me with her economical ways;

used to make me eat up the vile black potatoes, and what was called *hashed mutton,* which in fact was nothing more or less than a conglomerate of the fatty remnants of the past week. We were regularly regaled with this dish every Thursday. After I had left school some two years, one of Mr. Seyer's family told me that I had been a great favourite of his; certainly I should never have discovered his liking. I am sure the feeling was not shared by the French dancing master, for he expelled me from his class, and I think not without good reason. All the boys despised dancing, and at that time we had an inveterate hatred and contempt for all Frenchmen. Never shall I forget our boyish feeling of patriotism, the overweening self-confidence that we felt in our own prowess. I believe each boy in the school thought he was a match for six foreigners. On one occasion we got up a rebellion; we resolved to stand out for longer holidays. We arranged to barricade the schoolroom; we provided ourselves with muskets, and we had settled among ourselves what boys should station themselves at the windows, and when they were shot down which others should take their place. We were desperately earnest; not a boy but what was prepared to die at his post. However, the plot was discovered before it was ripe for execution;

the muskets were seized, the ringleaders expelled, others flogged; how I escaped I know not. One of these lads was afterwards midshipman in Nelson's squadron, and was cut down while bravely boarding an enemy's ship. But the most singular part of the business was that some of the Irish newspapers magnified our intended barring-out into an act of political disaffection, and reported in their columns that government was so unpopular that the British school boys were prepared to head a formidable riot of the townspeople. So much for the truth of history."

Andrew's appreciations of the beauties of the classical authors was intense, but his knowledge did not extend much beyond the limits of scholastic studies. He ever recurred with most pleasure to such portions as expressed sentiments which Homer breathes in the lines

> "Who dares think one thing, and another tell,
> My soul detests him as the gates of hell."

Of the Greek writers, Homer, Theocritus, and Thucydides were his favourites; of the Latins, Virgil and Horace. He used to speak of "the incomparable odes of Horace."

Mr. Crosse's father had been a friend of Franklin and Priestley; so we may conclude that he had some scientific tastes: but I never heard that he had directed his son's attention to these subjects. He may perhaps, almost unconsciously have become interested in electricity or chemistry from hearing his father's personal recollections of those two distinguished philosophers; but he himself gives quite another version of his first love for philosophy. "I had naturally," said Mr. Crosse, speaking of his schoolboy days, "a good appetite, and to this circumstance I attribute my scientific tendencies. When I was about twelve years of age, our drawing master lived some way from the school; the few boys who learnt took lessons at his house. I was not one of them, but I soon volunteered to become a pupil; for I discovered that there was a tavern not far from his residence, whose windows used to display most tempting joints of boiled and roast beef. I calculated that my drawing lessons would enable me to get out twice or three times a week to procure a good solid meal, which I stood much in need of. My father, who was much pleased at my own proposal to be instructed in drawing, readily consented to my becoming a pupil. Never shall I forget the lunches of nice boiled beef that the good old soul at the tavern used to cut off for me: she generally gave me more than my money's worth;

for she knew I was a schoolboy, and felt a pity for me. One day while discussing my beef, my eye fell upon a bill containing the syllabus of a course of lectures on Natural Science; the first of the series was on optics. I conceived a great wish to hear the lecturer; I asked and obtained permission of Mr. Seyer, to subscribe to the course. The second course was on electricity; my future tastes were decided."

It appears that all great minds peculiarly feel the shortness of human life: it is not the fear of death, but they are oppressed with the sense that time — that *no time* is enough for them. I have heard Mr. Crosse say: "I remember as if it was only yesterday, standing near the wall of our playground alone, and in a reflective mood, (I was often so, when not actively engaged in sports or mischief,) I thought to myself, suppose I was sure of living a hundred years, it would appear to me as nothing, so soon would it pass away; this feeling made me profoundly melancholy. But in general I was a very happy boy, careless, and extravagantly fond of fun. When I returned home for the holidays, I was made to read from the Greek three hours a day to my father, who was very strict. For my own amusement I had read whole volumes of the Philosophical Transactions, devoured Fielding, laughed at his Trullabeer, and formed an undying affection for Parson Adams. Voyages and travels were my delight. I liked Dean Swift's wit amazingly. I hated the pomposity of Richardson, and much preferred the broad coarse humour of Fielding. 'Sir Charles Grandison' was my aversion, a stiff, unnatural, ridiculous fool; Johnson, too, I hated."

When Andrew was about sixteen, he had the great misfortune to lose his father; had a few years more of life been granted to this judicious and affectionate parent, it would have been well for his son, now left to the charge of a mother whose only fault, but very grave fault, was an over indulgence of her favourite child. The natural impetuosity of his temper required a firmer hand than that of a woman to guide him in his early career. Of the four trustees appointed by Mr. Richard Crosse, namely, Mr. Hobhouse, Dr. Lovel, Dr. Jenkyns, and Mr. White, the two latter only lived to fulfil their trust, and they survived but a short while. At the end of Mr. Crosse's will he says, "May my children imitate the virtues of *my* father, and avoid the faults of *theirs*." These were the "good old times" when the honour paid to father and mother "made their days long in the land."

After his parent's death, Andrew returned to school. His love of electrical science increased.

His first experiment is thus detailed by John Jenkyns, Esq., his early friend and schoolfellow, in a letter to myself.

"* * * I dare say he has mentioned to you *our* first *joint* attempt in the science of electricity, and the wonderment occasioned to a circle of school boys by giving them a shock with a *Leyden phial* (a bottle which was sent to one of us with a dose of physic from the apothecary), charged by a broken glass of a barometer; the bottle was coated by myself, and I think he has told me not long since, that it was still in his possession. I fear that my knowledge did not go much further, but the world knows well what progress was made by my fellow-workman."

The following letter from Mr. Broderip addressed to Mr. Jenkyns well describes the schoolboy days of that knot of friends, of whom so few now remain.

"Gray's Inn,
"Feb. 20th, 1857.

"My dear J. J.

"You ask me for any reminiscences that may occur to me relative to our lamented school-fellow, Andrew Crosse, now alas! gone where so many whom I remember full of life and hope at the Fort, are—at rest, and happy, I hope and believe.

"When the electrical experiments began, Crosse was in the sixth form; and I was in the first and lowest, a child under seven years of age, and known to the 'big fellows' of the sixth only as 'a Black,' a thing to be bullied and belaboured.

"I well remember being brought up by Woodford—then I think in the fifth—to what I took for an old witch who was standing by an upright sort of doorless box; one of the 'presses,' as they were called, which stood in the hall. At the back of the box was a transparency representing a place which is said to be paved with good intentions; and before it, suspended and apparently dancing, pitchfork in hand, a frightful medieval devil. While I gazed in horror, a shock shot through my terrified frame, which I must have borne tolerably well, for I was afterward let into the secret, and assisted in bringing up other 'Blacks' to the scratch.

"The first electrifying machine was a broken barometer tube, rubbed with amalgam spread upon a piece of leather, which, if I recollect right, formed part of the lining of Ben Watkins's pumps. The Leyden vial was an apothecary's bottle, coated by yourself. The old witch was Andrew in a great coat with a pocket handkerchief tied under his chin and covering his head, and I am sorry to write that you were in attendance as a kind of familiar enjoying the terror and astonishment of the 'Black.'

"I well remember the dreadful battle between Crosse and Macdonald, and Seyer and his wife, for whom in revenge Crosse used to invent torments, imaginary of course, maliciously exhibiting poor black-eyed and swollen-nosed Crosse to us lower boys as an example.

"He was generous as well as brave; for though between

him and me as a 'Black' there was a great gulf fixed; he often gave me 'tuck,' out of those marvellous boxes of tartlets, game-pies, and cakes with which he was so liberally supplied from home.

* * * * * *

"Ever yours,
"W. J. BRODERIP.

"The devil, I afterward found was suspended by a single human hair, which was invisible to the frightened spectator."

After this first experiment the young electrician's enthusiasm increased, for one half-year he saved up every farthing of pocket money, manfully passed by the tavern, from whence the joints of boiled beef looked temptingly, and, hungry and half-starved as he was, kept his money.

He has often pointed out to me the shop on the Parade, Bristol, where he purchased his first piece of apparatus. The following letter I found carefully preserved amongst some old relics of his mother's.

"October 8th, 1801.

"Dear Mama,

"I received your letter the day before yesterday, and have taken this opportunity to answer it. As to my cylinder, &c., as they were not of so much value, I am consoled about their loss; but I have good news to tell you; there is peace. I suppose you have heard it before, but the *articles* are not yet signed; when they are there will be illuminations, bonfires, fireworks, and all the rejoicings that can be invented. The inhabitants of Bristol are mad with joy. As to the book you sent me — it is not very entertaining, but I hope you will send me some other soon. I am in good health, and hope you and Saturn are. Last Tuesday the boys did English verses on the peace; the following is a copy of mine.

"*De adventante Pace.*"

(The lines are too long to insert; they indicate the poetic talent which was afterwards developed.)

"Love to all.

"I remain
"Your dutiful son,
"ANDREW CROSSE."

I suppose some accident had happened to his cylinder, for soon after this he writes home to his mother for five guineas to buy an electrical machine. This request is granted, and he once more goes to the philosophical instrument maker's. He had been reading some works on electricity, and had been interested in some experiments, I think by a person of the name of Nicholls. "When purchasing my machine," said Andrew, "I entered into a philosophical discussion with the old man who kept the shop. I told him I was very curious about some experiments performed by a namesake of his mentioned in 'Nicholson's Dictionary.' 'Oh!' said he, 'I am the person. Those experiments you speak of are mine.' 'What!' I exclaimed in perfect amazement, 'are you really the person mentioned in the book? How glad I am to talk with you.' I cannot describe the profound respect I felt for the old man, and with what intense interest I listened to all he said. That I should know a real electrician, a man whose name was printed in a book with other philosophers, seemed to me an epoch in my life."

In June, 1802, Andrew left Mr. Seyer's, after being head of the school for a considerable time. Immediately afterwards he matriculated at Oxford, and became an inmate of Brazennose College. This was a critical period in his life. He had always disliked wine; but Oxford fifty years ago, was a sad place of temptation for a young and inexperienced boy. "I always hated wine," he used to say, in speaking of those times; "but I had not the moral courage to refuse joining the parties which were made up by my companions." In his usual strong mode of expression, he wrote home to his mother, saying, "Oxford was a perfect hell upon earth." — "What chance," he observed in after life, "what chance is there for an unfortunate lad just come from school — launched into every species of extravagance — no one to watch or care for him — no guide? (I often saw my tutor carried off perfectly intoxicated.) Most likely he falls in with ill-advisers, — and no one feels any shame, but the false shame of doing right. I confess I was influenced too by the aristocratic sentiments of those around me. I was less liberal at this time than at any other of my life; it took some years to rub off the prejudices of class which I had acquired at Oxford. I remember once, we were reading Aristotle on Friendship — 'Don't you think this rather too romantic?' observed my tutor. 'Not at all, sir,' I replied: 'I think a man ought to make every sacrifice for his friend.' I saw a smile of derision pass round the room, and I said no more. I felt I was not understood. Ridicule is a terrible trial to the young, and nothing is more rare than moral courage."

Andrew Crosse was twenty-one on the 17th of June, 1805. A shadow was upon his house — his mother was lying ill. The bells of the village church rang out a merry peal to celebrate his majority, but he indignantly stopped them, and passionately asked how they dared to ring when his mother was ill. "What," said she, "shall I not hear the bells ring for the coming of age of my eldest son?" "No, mother; I will allow of no rejoicings while you are ill and suffering."

Never was a parent more beloved: he would, I believe, have given his life for her, but their parting was near at hand. Mrs. Crosse died on the 3rd of July. I found the following affecting observation, written by Andrew on the outside of a letter from his mother, dated May 28th, 1805: "The last letter I ever received from this best of parents, who died in the following July. No pen can describe my *misery*." And on the back of the letter is written, "I have lost a father, mother, sister, uncle, Mr. White, Dr Lovell (two of my best friends), and a most faithful and attached servant."

He was now left

> "Lord of himself, that heritage of woe."

CHAP. II.

HOME LIFE AND "THOUGHTS."

1805—1821.

It has rarely been the fate of any one to be so associated through a long life with a particular locality as was Andrew Crosse. Broomfield was not only his home, but his abiding-place. The face of nature was as the face of a friend; and there, on those wild Quantock hills, with solitude round about him, and the varying clouds above him, he sought and found sympathy under all the trials of life, and his were not a few.

About twelve months after his mother's death he lost his maternal uncle, Thomas Porter, Esq. of Rockbeare House, Devon. Everything contributed to throw him back upon himself; with few friends but those of his own age, and with no good advisers, it will not be wondered at that he soon got entangled in difficulties. He had no business habits, and was wanting altogether in common prudence;

he implicitly trusted, without discrimination, in all those around him. Of course he soon became a dupe to the dishonesty of some, and a victim of his own and others' mismanagement. He used to say, in recurring to this time, "If I were to write a book, its object should be to show the mistakes people make at the commencement of their career. A boy comes from school, or a young man from Oxford or Cambridge, with a taste for the classics, and perhaps some knowledge of mathematics; but what does he understand of the management of his property, the duties of a magistrate, or the ordinary business of life? Agents, ever ready to assist the inexperienced, transact the young landowner's business for him, and his estates soon become involved. After years of discomfort, this truth dawns upon him, that if a man wishes anything to be well done, he must do it himself, and so it goes on; each man buys his own experience, but sometimes it is bought too dear."

For some three or four years, Mr. Crosse's brother and sister resided with him at Fyne Court; their *ménage* was more noted for hospitality than for economy; but they appeared to have been pleasant, happy days, for I have heard him say that he could have lived on for ever as they were living then. About this time Mr. Crosse was a good deal in the society of Theodore Hook, both in Somersetshire and in London; he was with the latter when he played off many of his well-known practical jokes. On one occasion Hook was dining with Mr. Crosse at his London lodgings; the day was hot, and the windows were thrown open; they were a merry party, and were talking and laughing loudly, when some wag who was passing by threw a penny piece into the room, which fell on the table close by a quarter of lamb that the host was carving. "Ah, mint sauce is good with lamb," cried Theodore Hook. I remember hearing Mr. Crosse say that he was once at a party with Mr. Hook, when a Mr. Winter was announced, a well-known inspector of taxes. Hook immediately roared out, —

"Here comes Mr. Winter, inspector of taxes,
I'd advise ye to give him whate'er he axes,
I'd advise ye to give him, without any flummery,
For though his name's Winter, his actions are summary."

No one loved a joke more than Mr. Crosse, and his anecdotes of this period of his life were very amusing; but he never indulged in any jest which could wound the feelings of another. Theodore Hook, it seems, was not always so particular.

At this period of his life Mr. Crosse intended

becoming a member of the Bar, and I believe kept two or three terms, but he soon abandoned all thoughts of the law, and shortly reverted to studies more congenial to his tastes.

He became acquainted with Mr. Singer the electrician, and during the short life of that gentleman (he died at the age of thirty-two) they were frequent companions. Mr. Singer supplied him with his splendid cylindrical electrical machine and battery table, which contained fifty large Leyden jars. The friends spent many pleasant days at Broomfield, in working together at statical electricity. They used also to take long walks over the Quantock hills; for some time they went every evening to one particularly romantic spot, to see the sun set. Mr. Crosse used long after to point out to me the paths they had taken over the hills, and he would recall their conversations together with great satisfaction, the more so, I suspect, because his opportunities of intellectual intercourse were few and far between. About this time Mr. Crosse appears to have worked hard in scientific matters; he was making himself a good practical chemist; he was studying mineralogy, and he kept a journal of electrical experiments. His investigations were principally directed to testing the power of the machine, under different conditions, and ascertaining the equality of the charging power of the positive and negative conductors. The science has so much advanced that the train of research that was then novel and interesting would not deserve the same kind of attention. I will not, therefore, intrude any of this class of experiments upon the reader.

The tendency of Mr. Crosse's mind was to examine all matters from his own point of view; it was not that he thought his opinion superior to that of others, far from it, for he was a peculiarly humble-minded man; and as his greatest desire in life was the attainment of truth, he hailed all new discoveries with genuine satisfaction, whether they resulted from his own labours or those of others. This peculiarity of clearing for himself a path in pursuit of an idea induced great originality both in the manner in which he conducted his researches and in the discoveries which ensued. He never accepted anything as true without first proving it by experiment, unless, indeed, the rationale of the proposition was such that it could not be otherwise. But if *he* viewed the cause of phenomena as arising from certain *unacknowledged* laws, he would not abandon his position, or desist from inquiry, because all former experience and the received explanation were opposed to the results which followed his experiments. It may be objected that there was a want of arrangement in the generality of his experiments. The facts stood isolated, and often did not do the good service they might have done; it may be said that he did not pursue his discoveries to their further consequences; he may not have sufficiently linked together the truths which he had successfully laboured to find; he was like the first discoverer of a mine of precious stones, he knew the value and properties of the rough-looking mineral which he brought to light; but he left to others the duty of polishing and setting the rare gem in the regalia of science, or applying it to purposes of utility. This and much more may be said, and said truly; but it is easy for the critic to look back and analyse a character with microscopic acuteness. This is certain, we must accept each man as he is; one mind may be fitted for analysis, another for synthesis; another without an atom of originality may ably generalise a confusion of facts,—all do good service, each in their way. We must not complain because Bacon was not Franklin, or Franklin a Bacon. There was an exceeding simplicity in Mr. Crosse's mode of experiment, and also in the occasions which led to the conception of an original idea. A stone perhaps picked up in his walks over the Quantock hills, suggested to his active mind an entirely new field of thought; he would then return to his laboratory, and set in action experiments to prove the correctness of his conclusions —and these were rarely unsuccessful; his imitations of nature were so close, simple, and obvious, that the process seemed like an intuition rather than an elaborated effort of the reasoning faculties. His motto was, "It is better to follow nature blindfold than art with both eyes open."

Mr. Crosse's first experiment on electro-crystallisation was about the year 1807. These investigations of the laws of nature arose from the following circumstance. In the parish of Broomfield there is a large fissure in a limestone rock, called Holwell Cavern; its roofs and sides are covered with arragonite in every possible variety of crystallisation. This romantic spot was often visited by Mr. Crosse, and suggested to him many poetic as well as philosophical thoughts. He pondered on the laws which regulate the growth of crystals; he could not believe that the starry emanations from centres were the effects of the mere mechanical dropping of water charged with carbonate of lime. Speaking of it he says, "When first I visited this cavern, I felt assured that I should sooner or later learn some new principle from an examination of its interesting crystallisation. I felt convinced at an early period that the formation and constant growth of the crystalline matter which lined the roof of this cave was caused by some peculiar upward attraction; and, reasoning more on the subject, I felt assured that it was electric attraction. I brought away some water from Holwell Cave, and filling a tumbler with it exposed it to the action of a small voltaic battery excited by water alone. The opposite poles of the battery were connected with the Holwell water by platinum wires, let fall at opposite sides of the tumbler. An electric action immediately took place, which continued for nine days; but not finding any formation upon either of the wires, I was about to take abroad the whole apparatus, when at that precise moment a party of friends called, and remained some time. This most fortunate delay prevented the removal of the apparatus till the next or tenth day, when I went for the purpose of so doing; the sun was shining brightly, and I plainly observed some sparkling crystals upon the negative platinum wire, which proved to be carbonate of lime, attracted from the mineral waters by the electric action." Mr. Crosse afterwards repeated this experiment in the dark, and he succeeded in getting crystals formed on the sixth day.

The following is a poetical description of Holwell Cave, extracted from a poem which Mr. Crosse wrote, entitled "My Native Hills": —

"Now pierce the hill's steep side, where dark as night
Holwell's rude cavern claims the torch's light;
Where, breathless, dank, the fissure cleaves in twain
Th' unchisell'd rock which threats to close again,
And swallow in its adamantine jaws
The bold explorer of creation's laws.
Boldness repaid (though rude the path) full well;
That entrance winds to nature's choicest cell.
Fancy may sport; but should her wildest pow'r,
Deluge some sea-girt cliff with ocean's show'r,
Then, while the spray flies feather'd by the blast,
Fix it in stiffen'd ice acutely fast;
Or should her art the shadowy deep explore,
And raise some corall'd grotto to the shore,
Such as might shield from Neptune's rude alarms
Fair Galatea tempted by its charms;
Nor wreathed icicle nor coral pale
Could mate the texture of that broider'd veil
Which nature in fantastic freak has thrown
In snow-like moss upon the rugged stone,
From which a host of vivid beauties rise
In unimagin'd forms to lure the eyes.
No two the same, yet each such art displays
That singly seen 'twould be unmatch'd in praise,
The needled tulip glittering on its stem,
The fair though Lilliputian diadem,

Bright groves of plants which every flower adorns,
And strange-form'd heads which rear their stranger horns,
Hands this outspread, that close which seems to hold
A fairy sceptre far outvying gold,
Hillocks of velvet lustre fring'd with brake,
And the cold crystals of the rigid snake;
Long waves in ribs across the ceiling thrown,
Which, though inverted, form a sea of stone;
Here mineral firs, whose downy foliage shines,
And feathery grass with brilliancy combines:
There groups of monsters arm'd with sparry claws,
Translucent sheets with edges jagg'd like saws,
Minutest fibres, some as tangled hair,
Some as sea-weed, beyond description fair;
All these and more the purest radiance fling,
Nor brush can paint, nor bard their wonders sing.
Yet not a breath at hand, nor distant sound,
Nor insect's hum disturbs the calm around.
Silence and sleep, and starless, rayless night
Here claim unquestion'd an eternal right.
The sheep's hoarse bleating and its tinkling bell
Pierce not the chasm nor disenchant the spell.
The peasant's whistle, and the watch-dog's bark,
The raven's croak, the rapture of the lark,
Die on their passage ere they reach the gloom
Or wake the echoes of the mineral tomb.
Here, whilst new realms arise and old decay,
And centuries of crime are swept away,
The night-born filagree of ages gone
Fenc'd from all living gaze creeps slowly on.
Pendant from arching roof the drops concrete

'Till the rude floor the growing crystals meet,
And arborescent shoots their branches twine,
As the soft tendrils of the curling vine,
The dazzling whiteness of whose stems might vie
With drifted snows that on the mountains lie."

I suspect that a variety of family and other circumstances contributed to distract his attention from the pursuit of this branch of science; at all events, there are no more records of what he did at this period, and I have reason to believe that he did not take up the subject again till some years after.

In 1809 Andrew Crosse married Mary Anne, daughter of Captain Hamilton; in the succeeding ten years seven children were born to him, three of whom died in childhood. These deaths, together with much family illness and many anxieties, early threw a shade over Mr. Crosse's life; he was one of those almost morbidly anxious persons who feel and suffer all things intensely. I have been unable to get any memorials of the first years of his married life; but from some records of his feelings, found in his own handwriting, he must have been much attached to his first wife. And of his fondness for his children in early life all bear witness who remember him in those days. Often and often has he been known to walk into Taunton, in the dead of a winter's night, to procure medical advice for his sick children, thinking, in his affectionate solicitude, that no one could do the errand as well as himself.

As Richard Crosse, his only brother, will be often spoken of in these pages, I will not omit the following characteristic announcement of his marriage. He, too, was clever in no common degree; and, if we take Andrew's estimate of him, he was, though totally unknown to the world, a profoundly wise man, to whom few were equal in power of intellect.

"Park Cottage,
"April 10th, 1813.

"My dear Brother,

"On the 6th of April I was married to a young woman that I have the highest opinion of; as to learning mathematics, she will learn anything to please me. I assure you I feel happier than ever I did in my life. At present I am quite unsettled and in confusion, not ever being used to domestic affairs. When I am settled I mean to attack mathematics with all possible vigour, and also to pay a little attention to music, but then I mean to adopt the new plan; that is to say, that the same alphabet shall be used to every clef, also that time shall be denoted according to Dr. Crotch's plan; it is, instead of saying Largo, Allegro, &c., it shall be denoted by saying 'Pendulum in inches.' But they must be new inches, or else it cannot be universal. And if I can get the universal concert pitch, I think the plan will be complete. I have got the woman who keeps the Globe Inn in Stowey to sell by the new measures. As for the old system, I mean to abandon it entirely. Indeed no person can be a friend to the sciences who adopts it; as for my part, I shall do my duty towards it, and endeavour to introduce it into some creek or corner if not universally. I have one comfort that I have an active agent in the East, that is, old Staines. If there is ever a country election again I will not give any one my vote unless they will promise to bring forward the new system to Parliament. Remember me kindly to Mrs. Crosse and Mrs. Hamilton, and believe me,

"Ever your most affectionate Brother,

"RICHARD CROSSE."

Mr. Richard Crosse was an ardent supporter of the decimal system. He literally carried out his principles in all things. His clocks were divided into ten hours; his weights and measures were all on the metrical system. And so strictly mathematical was he in all his ideas, that when he built a house it was in the form of a double cube, each room being also mathematically proportioned. Mr. Crosse used to call his brother the three M's; for his favourite studies were music, mathematics, and metaphysics; in the latter subject he was deeply read. And his mental influence upon Andrew was immense, reminding one of the effects of William Humboldt's studies upon his brother Alexander.

Mr. Crosse often congratulated himself on having had such a companion and friend as his metaphysical brother, for the tendency of his mind was eminently useful in counteracting the materialistic effect which sometimes follows an exclusive devotion to the physical sciences.

There are few incidents to record of this period of Mr. Crosse's life; he lived quietly at Broomfield, occupied in his duties of husband, father, landlord, and magistrate. Some extracts from letters written to his dear friend and schoolfellow, John Kenyon, Esq. (himself the intimate friend of Wordsworth, Coleridge, Southey, and other distinguished men of the time), will, perhaps, be the best transcript of these days.

"Broomfield, July 23rd, 1815.

"Dear Kenyon,

"It is two months since I received your letter, but no want of inclination has delayed me from answering it before; the truth is, I have been more busy than ever *winding up* my last business with workmen of all descriptions. I have now totally finished with the carpenter, mason, smith, plumber and glazier on my house and premises, having left nothing, not even a hinge, or a handle of a drawer, but what is completely in order. * * What think you of Bonaparte surrendering himself? What will they do with him? France is in a dreadful situation. I could poetise, but have no time, and am tired of hack subjects. I like all your verses, but I infinitely prefer those on Broomfield Churchyard: they are simple, melancholy, to the purpose. I should like to see Bonaparte: if he lands at Plymouth I shall endeavour to get a peep at him. It must be gratifying to rest one's eyes upon a man who will live to the end of time, whose name will be in the mouths of all nations, when our bodies are mouldered to dust. Excepting him, and Wellington, I have no great ambition to see the continental kings, who will only fill up the intervals of history, like notches in a prisoner's stick. I am sorry to send you a double letter about nothing: indeed, should I write a quire, I should find it too small for what I may wish to say to you. Success attend the man who invented the mail coach! * * * I shall be delighted to hear from you on your travels. However, I should scarcely recommend Switzerland at present; it is in a very unquiet state.* * It appears to me, from the observations in the 'Courier,' that our Government is a good deal dissatisfied with the king of France's choice of ministers. I should not wonder if they pulled him down from the throne again. He is a mere idiot — a child in leading strings. We shall, I suppose, keep Bonaparte safe *in terrorem* to the Bourbons.

"Yours, &c.,

"ANDREW CROSSE."

He carried out his intention of seeing Napoleon, for on a scrap of paper is the following: —

"1815.

"I left home on Wednesday morning about eight o'clock, reached Exeter on my mare about four o'clock, left Exeter ten o'clock same night, reached Plymouth Dock at half after five Thursday morning, saw Bonaparte about six o'clock same night. He started in the Bellerophon, without signal, at about one o'clock, Friday, August 4th, 1815, was met by a sloop-of-war, who exchanged signals with the Bellerophon; followed by the Eurotus frigate with French officers of rank aboard, also by the —— and Glasgow frigate and Mackerel schooner. The Bellerophon bore eastward, and I followed her some hours; was on the water from twelve to two; sailed ten miles from dock, and about twenty-five or thirty in all. Saw Bonaparte between decks, afterwards, in the cabin window, the curtain of which was drawn by a French lady; got within thirty yards of him, and was told by Captain Maitland to keep off."

He writes again to his friend: —

"Broomfield, 1817.

"Dear Kenyon,

"* * * The idea which you mention as prevalent on the Continent, viz. of a disbelief of any disinterested virtue, is beyond everything repugnant to a generous mind, and most certainly untrue. * * My philosophical friend Singer has had the misfortune to break a blood-vessel in the lungs. I fear his mortal career must soon terminate — he whom I recollect a twelvemonth past enjoying the sound health of a temperate philosopher, with a body more than commonly muscular and unimpaired by any excess, with a calm, vigorous, discerning mind. Now he is forbidden to read and scarcely to think; he who made the best use of his time of anyone I ever knew. Sic transit gloria mundi."

In another letter to Mr. Kenyon, dated a few weeks later, he writes: —

"Poor Singer died yesterday. I truly lament his loss, but he died like a real philosopher.

"'*Felix qui potuit rerum cognoscere causas,*
Atque metus omnes et inexorabile fatum
Subjecit pedibus strepitumque Acherontis avari.'

"No man, whatever his rank, wealth, strength of mind and body, can say, in the vigour of his life and full enjoyment of all these blessings, I am happy. Can a person be happy who sits on the verge of a precipice? and yet a fall down a precipice would be Paradise compared with the misery which may await us in a future state. But without looking so far — which is, however, a near prospect to the youngest of us, — can we be happy whose pleasures are suspended by a hair? The loss of a near relative or friend, of health, &c., may plunge such a one in greater misery than any happiness he ever felt. If there was no future state, it would be beyond comparison better never to have been born. I who say this am not tormented by Blue Devils. I can be comparatively happy in all situations and places, and could live a thousand years as I now am without feeling *ennui;* but I know that were I to experience the loss of a child, or any other misfortune of that nature, I should suffer more pain than I ever felt pleasure. * * * Man is born for the good of man; and that person who spends his life in actively benefiting his fellow-creatures, and amuses himself in hours of relaxation with mental enjoyments, need not fear the awful hour of death. I must, however, check my moral strain, and inform you of the news of the country.* * * I am continuing my *electric poem* at intervals. Your account of the Italian scenery transports me to my old bench at Seyer's, where, being in expectation of our being called up by his thundering voice, I spent my time in classical dreams of unalloyed future happiness, fearing no enemy but Seyer, and believing firmly that those prodigies of perfection described in such glowing language by poets and historians were not unfrequently to be met with in real life. Yet I cannot say, '*Oh mihi prœludos,*' &c. I invariably feel melancholy when I look back on past times, read past events, or examine antiquities; not from a wish of recalling such times, but from a consciousness that all belonging to myself, my family, my friends, will shortly be swept away in the same manner, and form a portion of that past which it is permitted me for a short space of time to contemplate. The fine speech of Andromache always rises to my imagination, in which, when in the zenith of her husband's power and glory, she depicts the time when possibly she may be pointed at as a slave in Greece, condemned to draw water, and when the passenger might say,

> Ἑκτορος ἡδε γυνη, ὁς αριστευεσχε μαχεσθαι
> Τρωων ἱπποδαμων ὀτ' αρδ' Ἴλιον αμφεμαχοντο, &c.

"But it is of no use to lament what cannot be otherwise, and madness to set one's heart upon what is so transitory. I well remember when, at one-and-twenty, I said I should be miserable if I thought I should die without having seen Greece and Italy; and yet now I must sit down contented with only an historical acquaintance with those countries, having a thousand other subjects to attend to. Your account of Rome and Venice agrees with my own preconception of them; the one engendering an infinity of thoughts,—Tarquin banished, the entry of Coriolanus, the consuls, fasces, &c.; the entry of Brennus with fire and sword, Marcius, Scipio, Cæsar, &c. &c., all passed away like the painted slides of a magic lantern. The other striking the sight with a beautiful vision of palaces rising out of the sea. The one like Lear, magnificent in hoary ruin; the other like a prince arrayed in modern splendour. * * *

"My brother is full of metaphysics and Kant. He certainly argues more closely than anyone I ever conversed with: like Dr. Johnson, 'a fallacy won't stand before him.'

* * * * * *

"Yours ever sincerely,

"ANDREW CROSSE."

In the year 1816-17 Mr. Crosse appears to have been carrying on his scientific pursuits with ardour.

Some years previously he had erected an extensive apparatus for examining the electricity of the atmosphere. The first time Mr. Crosse's name was mentioned in print in connexion with science was in "Singer's Elements of Electricity and Electro-Chemistry," published in 1814. From the pages of this work I give the following extracts : —

"My friend Andrew Crosse, Esq., of Broomfield, near Taunton, a most active and intelligent electrician, has lately made very numerous observations with a remarkably extensive atmospherical conductor, consisting of copper wire. * * * The most unwearied exertion has been employed to give unexampled extent and perfection to this apparatus. The insulated wire has been extended to the extraordinary length of one mile and a quarter * * * It has now been deemed expedient to shorten it to 1800 feet. * * * A wire of this kind has been kept strained for eighteen months without injury, and from observation of its indications, and those obtained in other experiments of less duration, the following deductions have been made : —

"1st. In the usual state of the atmosphere, its electricity is invariably positive.

"2nd. Fogs, rain, snow, hail, and sleet produce alterations of the electric state of the wire. It is usually negative when they first appear, but oftentimes changes to positive, increasing gradually in strength, and then gradually decreasing, and changing its quality every three or four minutes. These phenomena are so constant, that whenever the negative electricity is observed in the apparatus, it is considered as certain there is either rain, snow, hail, or mist in its immediate neighbourhood, or that a thunder-cloud is near.

"3rd. The approach of a charged cloud produces sometimes positive and at others negative signs at first. * * * During this display of electric power, so awful to an ordinary observer, the electrician sits quietly in front of the apparatus, conducts the lightning in any required direction, and employs it to fuse wires, decompose fluids, or fire inflammable substances; and when the effects are too powerful, he connects the insulated wire with the ground, and transmits the accumulated electricity with silence and with safety.

"4th. A driving fog or smart rain frequently electrifies the apparatus nearly to the same extent as a thunder-cloud, with similar changes.

"5th. In cloudy weather, weak positive electricity usually prevails; if rain falls, it generally changes to negative, but the positive state is resumed when the rain ceases.

"6. In clear frosty weather the positive electricity is stronger than in a fine summer's day; and the intensity of the electrical signs at different seasons is expressed in *descending* order in the following list, commencing with those whose effects are most considerable.

"1st. During the occurrence of regular thunder-clouds.

"2nd. A driving fog accompanied by small rain.

"3rd. A fall of snow, or a brisk hail-storm.

"4th. A smart shower, especially on a hot day.

"5th. Hot weather succeeding a series of wet days.

"6th. Wet weather following a series of dry days.

"7th. Clear frosty weather, by day or night.

"8th. Clear warm summer weather.

"9th. A sky obscured by clouds.

"10th. A mackerel back or mottled sky.

"11th. Sultry weather, the sky covered with light hazy clouds.

"12th. A cold rainy night.

"To this may be added the least electrical of all — a peculiar state of the atmosphere which sometimes occurs during the prevalence of north-easterly winds: it is characterised as particularly unhealthy, and is remarkable in producing a sensation of dryness or extreme cold, which is not accompanied by a corresponding depression of the thermometer.

"The usual positive electricity is weakest during the night. It increases with the sunrise, decreases towards the middle of the day, and increasing as the sun declines, it then again diminishes and remains weak through the night.

"This fact is one of the most instructive resulting from these observations, and is confirmed by most of the regular experiments on atmospherical electricity that have been made. It clearly proves that the electricity of the atmosphere is influenced by the same causes that promote the equal distribution of moisture."

The exploring wire was brought into the electrical room at Fyne Court, and the effects produced under certain states of the atmosphere, were little short of terrific; at least as far as noise and brilliant light was concerned, but harmless and obedient under the control of the practised hand. In Dr. Noad's "Manual of Electricity," he thus describes the power of the exploring wire.

"The electrical battery employed by Mr. Crosse, consists of fifty jars, containing seventy-three square feet of coated surface: to charge it requires 230 vigorous turns of the wheel of a twenty-inch cylinder electrical machine; nevertheless, with about one third of a mile of wire, Mr. Crosse has frequently collected sufficient electricity to charge and discharge this battery twenty times in a minute, accompanied by reports almost as loud as those of a cannon. The battery is charged through the medium of a large brass ball, suspended from the ceiling immediately over it, and conducted, by means of a long wire, with the conductor in the gallery; this ball is raised from and let down to the battery by means of a long silk cord, passing over a pulley in the ceiling; and thus this extraordinary electrician, while sitting calmly at his study-table, views with philosophic satisfaction the wonderful powers of this fearful agent, over which he possesses entire control, directing it at his will; and with a single motion of his hand banishing it instantaneously from his presence."

Mr. Crosse himself thus describes the effect of a dense fog on his electrical wires. "I was sitting," he says "in my scientific room, on a dark November day, during a very dense driving fog and rain, which had prevailed for many hours, sweeping over the earth, impelled by a south-west wind. The mercury in the barometer was low, and the thermometer indicated a low temperature. I had at this time 1600 feet of wire insulated, which, crossing two small valleys brought the electric fluid into my room. There were four insulators, and each of them was streaming with wet, from the effects of the driving fog. From about eight in the morning until four o'clock in the afternoon, not the least appearance of electricity was visible at the atmospheric conductor, even by the most careful application of the condenser and multiplier; indeed, so effectually did the exploring wire conduct away the electricity which was communicated to it, that when it was connected, by means of a copper wire, with the prime conductor of my eighteen-inch cylinder in high action, and a gold leaf electrometer placed in contact with the connecting wire, not the slightest effect was produced upon the gold leaves. Having given up the trial of further experiments upon it, I took a book, and occupied myself with reading, leaving by chance the receiving ball at upwards of an inch distance from the ball in the atmospheric conductor. About four o'clock in the afternoon, whilst I was still reading, I suddenly heard a very strong explosion between the two balls, and shortly after many more took place, until they became one uninterrupted stream of explosions, which died away and recommenced with the opposite electricity in equal violence. The stream of fire was too vivid to look at for any length of time, and the effect was most splendid, and continued without intermission, save that occasioned by the interchange of electricities, *for upwards of five hours,* and then ceased totally. During the whole day, and a great part of the succeeding night, there was no material change in the barometer, thermometer, hygrometer, or wind; nor did the driving fog and rain alter in its violence. The wind was not high, but blew steadily from the south-west. Had it not been for my exploring wire, I should not have had the slightest idea of such an electrical accumulation in the atmosphere: the least contact with the conductor would have *occasioned in-*

stant death, the stream of fluid far exceeding anything I ever witnessed, except during a thunder-storm. Had the insulators been dry, what would have been the effect? In every acre of fog there was enough of accumulated electricity to have destroyed every animal within that acre. How can this be accounted for? How much have we to learn before we can boast of understanding this intricate science!"

Mr. Crosse's electrical machine was frequently in requisition for medical purposes. The poor in the neighbourhood used to go to Fyne Court to be electrified for paralysis and rheumatism, and in almost every case the effect was highly beneficial. I remember hearing of a farmer, a man upwards of sixty, who was paralysed on the left side, and had besides a distressing complaint of the salivary glands. At first, when he went to Mr. Crosse, it was with difficulty that he could be assisted out of his gig; after being electrified twice a week for six weeks, he was so much better, that he could walk to Fyne Court, and the complaint in the throat was entirely removed.

Voltaic electricity occupied Mr. Crosse's attention at this period; for there are memoranda of his, bearing date 1817, wherein he describes his experiments on crystalline formations.

By his close and ingenious imitation of Nature's arrangements, together with the application of electricity, he formed various metallic and earthy crystals, which had never before been formed by art, and in some instances he produced substances that have never been found combined in nature. I don't think Mr. Crosse was at all impressed with the importance and interest attached to this discovery of these laws of nature. He had no scientific friends with whom to communicate; he lived at Broomfield in perfect intellectual isolation: this circumstance and some others contributed to develope in him the poetic tendencies which were with him from his childhood. Not only did he love the hills, the trees, the streams, the flowers, but the wild animals that found an asylum in his grounds became objects of his affection; he used to say, "I think I could tame every animal in creation, except my own species." He had many singular adventures with animals.

His sympathy with them, and his feelings generally, are well portrayed in the following poetical pieces, which date about this time.

"LINES

ON A RED DEER TURNED OUT BEFORE THE STAG HOUNDS, ON BROOMFIELD HILL.

"Away! away! to the hills away!
The hounds are on thy trail!
Away! away! at a breath's delay
Thy blood shall stain the vale!

"Away! away! beleaguered thing!
O'er bank, and bog, and brake!
Thy foes' fierce shouts behind thee ring;
Thy life — it is the stake!

"Trust to thy speed thy friend to prove,
But trust not man; for he
Craves mercy from the Power above,
But none he shows to thee!

"Trust to the wilds, the floods, the wave,
The precipice, the plain:
If such as these may fail to save,
Thy trusting is in vain!

"Strain every nerve, exert thy skill,
And thread the tangled maze;
Baffle pursuit in pool or rill,
Each step on land betrays!

" And now, if close thy foes are found,
Then bravely turn and die!
The freeborn gales that float around
Shall waft thy final sigh!"

"MY DOG.

" I'LL lock thee, love, within my breast,
And hoard thee for a happier state;
There 's little here to be caressed,
When nought but coldness finds a mate!

" I'll hide thee — no, my dog shall share
What human heart could ne'er repay;
What friends or condescending fair
May first entice and then betray!

" But thou! untutored, soulless brute,
Whose life, men fancy, is thine all,
Would'st give that life, though thou art mute,
To save thy viler master's fall!

" Then let us boast our high degree,
And preach of what we are to win,
Better thy kind mortality
Than our immortalising sin!"

"TO THE FOXGLOVE.

"Dearest of all the flowers that grace
The heathy hill or tangled brake,
That to the sun turn glowing face,
Or bow them o'er the shady lake!

"That climb some friendly stem to show
Their beauties to the wanderer's eye,
Or, dimly seen, their fragrance throw,
Sweet duty! ere they droop and die!

"Dearer than all the rainbow hues
Or tropic tints on shrub or tree,
Or sketched by pencil or by muse,
Art thou, surpassing flower, to me!

"Surpassing! ay, thou dost surpass!
For in thy spotted bells I view,
Mirrored as in some fairy glass,
Past days my kindling thoughts renew!

"Thou seem'st to smile with such intent,
That as on thee abstract I gaze,
I fain would lose what Fate has sent,
And dream the dreams of bygone days!

"The rose its grateful scent may yield,
But soulless is that scent to me!
The lily paint the varied field!
What are the lily's charms to thee?

"Our earliest thoughts are holiest ties:
I loved thee tall in purple bloom;
Thy coleurs cheered my infant eyes,
And thou may'st blossom near my tomb!

"Then let them vaunt in song elate
Their spicy grove or treasured vine;
I care not for the vulgar bait,
The graceful foxglove shall be mine!"

SPRING.

"'Tristitiam et metus tradam protervis ventis.'
HOR.

"WHOE'ER owns a heart, now let it expand!
All hail to the blossoms that bloom o'er the land!
All hail to the fragrance that floats on the gale!
All hail to the songster that skims through the vale!

"Off, cares and vexations! I'll fling ye away!
And though ye'll come back, ye shall not come to-day.
In the meads will I revel. What Fate may assign
I toss to the winds: this day shall be mine."

"TO THE WINDS.

"SAUCY Winds, come do your duty;
Play around my buoyant feet;
Soft or blustering as may suit ye,
I shall not from your force retreat!

"Sport in gusts, howe'er disdainful,
Welcome are those gusts to me;
The winds of heaven can ne'er be painful,
Come they from the land or sea.

"Ye sweep perchance from craggy mountain;
Thyme and wild mint flourish there!
Ye cross in spite the ruffled fountain,
But waft its coolness in your air.

"Sigh ye may in sullen sadness,
Still those sighs are made to heal;
Or rage aloud in phrensied madness,
Your wrath does but new charms reveal!

"What were the trees were ye to slumber,
Were there no gales their limbs to sway?
What were the flowers, their hues or number,
Did not ye give their scent away?

"What were the waves of listless Ocean,
Did not ye break its glassy sleep,
And rouse the bard to stern emotion,
As o'er the rocks the breakers leap?

"Nature ne'er meant this vast creation
To lie one dull lethargic whole,
But, mistress of her great vocation,
Gave to the mass a glowing soul.

"Thus from their deep recesses beaming,
Sprung life, and light, and joy, to bless;
And billowy waves, and waters streaming,
Their mighty Maker's hand confess.

"And you, ye Winds, howe'er aspiring,
Slow in your course, I fain would find,
When brighter worlds my soul desiring,
Leaves its dull clay a wreck behind."

"GOOD AND ILL.

"Lament not, oh, lament not,
Ye who of ill complain!
For all your tears prevent not
The sure approach of pain!

"Go skim the troubled waters!
Flit o'er the earth's dark breast!
Where are her sons and daughters,
Who tell you they are blest?

"Or should some bright delusion
Just set their fears asleep,
They soon, to their confusion,
Find all are born to weep!

"Yet hope inflates our pillows,
And gambols in life's dream;
It gilds the rudest billows,
And prattles in the stream.

"It crests the savage mountain,
Its chorus rends the cloud;
Its sun plays o'er the fountain,
Or stoops beneath the shroud.

"It lends the gales their sweetness,
The groves their rich array;
Each coming joy its fleetness,
And bids each passing stay.

"It clothes the stars with power
To shed a purer light;
It sparkles in the shower,
And whispers in the night.

"Without its aid creation
The aching sight would tire;
The tidings of salvation
Awake no holy fire.

"Thus good and ill before ye
Your Maker deigns to bring;
He blends the shame with glory,
The winter with the spring.

"The ill to show your weakness,
The good to prove his care;
'T is that may teach you meekness,
And this to shun despair.

"Then praise to Thee, Most Holy,
Though good or ill betide!
Oh make thy people lowly,
And cast away our pride!"

"THE CHAMBER CLOCK.'

"List to the chamber clock,
As it measures the time with regular knock!
As the pendulum sways from side to side,
It beats to the fall of mortal pride.

"Beat! beat! the Spring draws near,
And promises fair for a future year;
The Summer awakes — it glows — it flies,
Then quick gives place to autumnal dyes.

"But withering Winter triumphs in turn,
And scowls o'er Nature's funeral urn.
Beat! beat! in the glare of day,
When the dream of life is bright and gay.

"Beat! beat! in the calm of night,
When darkness shrouds the waker's sight;
When memory views with unclouded eye
The faces of friends that are long gone by.

"Beat! beat! in love or strife —
Every beat is a slice of life,
And when the listener 's dead and gone,
Another shall hear the clock beat on!'

"LOVE AND HATE.

" There is a love which none can teach,
 Deep as beneath the wave
Sleeps the fair gem, no ray can reach,
 Within its ocean grave.

" There is a hate of such a kind
 Nought earthly can assuage,
And Death himself must weep to find
 Its force surpass his rage.

" To few that mighty love has sped,
 To few that hate so curst;
And he who fain the last would wed,
 Must revel in the first.

" Must revel — ay, the cup must quaff
 Till not a drop be there,
Then see his kindness as the chaff
 Tost to the jeering air.

" Then will he hate; and every vein
 Where love once held her throne
Will wake to never-dying pain,
 To a hell before unknown."

"FRIENDSHIP.

"Ah ! 't is not the time when your friend is in trouble,
Though fragile and faulty that friend has been found;
Ah! 't is not the time to prove friendship a bubble,
And with the base herd tread him down to the ground!

"If his fate has been just, is it well to upbraid him,
And then to the Great One for pardon to kneel?
To expect that the God whom he calls on to aid him,
To thee will be tender, to him will be steel?

"If his fate has been hard, is it right to deny him
The kindness you once in prosperity shared?
Or, that which a savage would shrink from, to fly him,
As if at his presence polluted or scared?

"Yet such is the world — all religious and saintly
When religion 's the passage to tinsel and pride,
But retailers of doctrines which sound somewhat quaintly
When the word by the deed is expressly belied!"

"TO THE BELLS WHICH RING OUT THE OLD YEAR AND RING IN THE NEW.

'Ye wanton sounds, that dare invade
With reckless glee the midnight hour,
When all is wrapt in solemn shade,
And Sleep would fain assert her power:

"Why burst ye on the stillness round
With flighty taunt and accents new,
Like idle jest careless to wound
When grief demands its homage due?

"Is it that your triumphant strains
Laugh that another year is fled,
And that, alas! one less remains
Ere earth shall close upon its dead?

"Is it to vaunt that time gone by
Has swept all wintry storms away,
That grief no more shall raise a sigh,
And phantom hopes no more betray

"By strange though vain attempt to blind
Us, too well school'd in Nature's lore?
To leave unwept all pain behind,
And view all pleasure placed before?

"To tell — what every gale denies
Which strews the blossoms on the plain,
Or when, more wrath, the angry skies
Hurl fierce defiance o'er the main?

"Would ye disturb the silent crowds
Whose dust forgotten sleeps below,
And wrench them from their mouldering shrouds
For this sad world of sin and woe?

"Again to rouse the victor's ire,
To wrap new towns in sheets of flame,
Again command the poet's lyre
To hymn those deeds of blood and shame?

"Would ye call back to mortal strife
The maniac's scream, the felon's gloom,
The unutterable woes of life,
Again to sink into the tomb?

"The bird that wails from its lone dell,
The moon's ray sleeping o'er the lake,
The rippling stream, all form a spell
That mortal music scarce may break.

"And yet I still would hail your song,
Which those I loved once heard with glee;
And while its swell the winds prolong,
Would dream of pleasures ne'er to be!

" Would dream — alas! poor human dreams,
Like your sweet tones, soon fade in air,
And spring's green buds, and summer gleams,
Must sink in death, however fair!

" Still, e'en on earth all is not lost,
Though roses droop, and joys depart;
For sharpest thorns, and fancies crossed
Shall cease alike to yield their smart:

" And Happiness, though swift her wing,
Shall drag afar her sister Pain;
For neither shall a pæan sing,
Nor boast a solitary reign.

" Then sport, ye bells, in thoughtless peal;
For as your notes together blend,
Their echoes thrill with woe and weal,
Linked arm in arm till time shall end!"

" SPRING.

" So bounteously does Spring, mankind to bless,
Pour forth her charms, she dies in the excess."

"THE THREE TRENCHES.

"THREE circling trenches round my heart I throw,
To keep at bay each intermeddling foe.
Within the first the world may enter free,
Whate'er their sect, opinion, or degree;
Safe o'er the next, I greet a fair array,
Serenely smiling as a summer's day;
To pass the third, alas! how few contrive!
And of those dearest few, how few survive!"

"Broomfield, March 10th, 1822.

"My dear Kenyon,

"Some eleven months since, I had, by dint of coaxing and tormenting, prevailed upon a roving unsettled friend of mine to spend some two or three of his uncertain days at Broomfield, and during his short stay I was under no small anxiety concerning his health, which, according to him, was on the decline. He took me with him to look at some houses in the neighbourhood, which he might have taken a lease of, but for the very precarious, delicate, and uncertain state of his health. He was in an atrophy—spasmodic, intermittent, inflammatory fever—or some such doleful distemper, accompanied with groanings, sighings, heart-throbbings, pulse-thumpings, soliloquisings, lamentations, and such a prolonged variety of symptoms, that I was fool enough to suppose him really ill, and his case so uncommon that it should be read at one of the Royal Society Meetings. He would, in fact, have puzzled heads much more accustomed to medicals than mine.

" 'Hard by yon wood, now smiling as in scorn,
Muttering his wayward fancies he would rove;
Now drooping, woful man, like one forlorn,
Or crazed with care, or cross'd in hopeless LOVE.'

"I wish you could have seen him, as on a sudden he turned round to me and said, 'Crosse, do feel my pulse, — what do you think of me?' This he repeated hourly. I endeavoured to comfort him, but to little purpose. I heard a few months afterwards, that this very identical invalid was — ye gods! — married; fat, sleek, *well conditioned*, sound in wind and *limb, bereft of angles*, acids, and troubles, and become a *sober, steady, hearty* consumer of the good things of this life, instead of being himself a prey to pining disease or wandering melancholy. I know not whether any of this relates to your present condition; but this I know, that I am heartily glad to find that you have been so fortunate in your marriage, and wish you many years of happiness. * * * My eldest boy has pleased his master by a copy of Latin verses; my other boys are going on well; Mrs. Crosse's health is much the same. My time is so much occupied, that poetry must wait. I have just scrawled a Sapphic ode, *Romanus*, applicable to the present state of England. * * * You mention geology in your letter. As to the Spaxton Cave, it is composed wholly of carbonate of lime, without a trace of fossil bones; but a few remains of antediluvian vegetation are impressed on some of the limestones in its vicinity. My studies, however, just now, are of a different character: I am reading the Classics with my two youngest boys, and am preparing an analysis of the Greek verbs. My eldest son is still at Coleridge's, where young St. Aubyn goes, as their parents did to Seyer's. How the years fly, and how we are all changed! a few more years, and nothing will remain of us but some chance fusty manuscripts, half worm-eaten, showing perhaps some remains of my iron-ribbed letters. * * The report of my being about to enter into holy orders is correct * * I fear I am *utterly unworthy* of taking upon me so sacred a function * * God bless you; and believe me, Kenyon,

"Yours most sincerely,

"ANDREW CROSSE."

About this time the mere business affairs of life pressed heavily upon Mr. Crosse: he was never a good manager, careless about money matters, and, though not in the least extravagant, he was often injudicious in his expenditure. He was never a rich man, considering his position; his father used to say that he could only leave him bread and cheese. The Crosses had been a wealthy family, and had possessed lands in various parishes in the west of Somersetshire; but, as Mr. Crosse often observed, "my family were learned and honourable men as long as I can look back; but they had the happy knack of turning a guinea into a shilling, and I have inherited that faculty pretty strongly." The necessary expenses of a large family, alterations in the house and grounds, rebuilding of farm premises, and those many drawbacks of landed property, all contributed to cripple his resources. He used often to repeat the saying of some one who says, "Land is principal without income, and the funds are income without principal." Amidst the petty annoyances which irritate, and the sorrows which overwhelm, Mr. Crosse still pursued the studies which were most congenial to his nature. Amongst his papers I found the commencement of a poem, in blank verse, on "Electricity," scraps of unfinished poetry on almost every imaginable subject, translations, essays and scientific notes. His handwriting was most peculiar and characteristic. To some persons it was not very legible, occasionally not very clear to himself: he satirised it in the following lines: —

"Belshazzar's soul was sore amazed
When on the mystic lines he gaz'd
Traced on his palace wall;

When Daniel rose, high gifted-seer,
The fatal prophecy made clear :
But Daniel, were he now alive,
In vain might scratch his head and strive
Thy meaning to explain ;
Thy mazy pen would bid defiance
To saints and devils sworn alliance,
All efforts would be vain."

But, in reality, his handwriting did not deserve this abuse. It was bold, vigorous, decided, and in a certain sense regular : it certainly bore out his own theory, that character can be told by the writing.

Amongst Mr. Crosse's poetry there is a fragment of some 250 lines, of which he observes in a note, "The lines arose from a real train of thought which occurred to me." I well remember his alluding to the circumstance. He had been to Plymouth to see a young friend off, who was going to India to join his regiment. Mr. Crosse had come back as far as Exeter, where he slept; he was very much fatigued, and felt low. He described that he had scarcely lain down upon his bed, when a sudden train of thought burst upon him with such intensity that it seemed almost like inspiration : he was not asleep, it was no dream; but yet in imagination he roamed over the universe; and beheld with the eye of fancy the unbounded glories of creation; it appeared to him, he said, as if the soul had quitted its prison of clay, and was free to reach the limits of space, or rather to annihilate space with the intensity of its perception. Centuries of time were condensed in those moments of ecstatic life, and Nature's laws seemed clear to the omniscience of its glance: a sense of blessedness sustained him — he *felt* immortal.

> "So thought I, in the dead of silent night,
> Till, with the flow of imagery tired,
> I sank resistless in the arms of sleep."

Thus ends the poetical description of his reverie, which reminds one of what Sir Humphry Davy describes that he felt while under the influence of nitrous oxide gas.

About this time, Mr. Crosse continued to write a great deal of poetry, which might almost be said to supply the place of an autobiography, it is such a full and perfect transcript of his feelings and opinions, his sorrows and aspirations. It has often been said that an author's life is best read in his works. The biographer's duty is to record events, which, after all, often stand apart from the inner life. The severe character of history, even though it be the history of a humble individual, does not permit the indulgence of metaphysical speculations; it can but record "noted action." The novelist has manifestly the advantage: he can, without departing from truth or propriety, reveal the inner life of his characters; he can minutely detail the development of their mental and moral growth.

Mr. Crosse used himself to observe, "What can we really know of the history of any man?" Some idea of this kind perhaps made Walter Savage Landor exclaim, "Fiction is more true than fact:" this paradox applied doubtless to what was true to human nature in contradistinction to historical truth, which, after all, is but the external semblance of unexpressed realities.

The following paper, written by Mr. Crosse for the occasion of an archæological meeting in the county, is inserted here, because it describes in graphic language the scenes in which he passed his life, and reveals the influence of natural objects and local associations upon his mind.

"WALKS ON THE QUANTOCK HILLS.

"The love of our country is one of the most prominent feelings incidental to man, and it seems that the inhabitants of wild and mountainous regions are more warmly attached to their partially cultivated deserts, than those who dwell in rich and luxuriant plains which yield to the cultivator a fifty-fold return for far less labour. Man is born to toil; nor can health of mind, or vigour of body be ever attained by those who live in fat indolence. Indeed, we commonly see that when Nature does much for the soil, she does proportionally less for the minds of its cultivators. Necessity is truly said to be 'the mother of invention,' and where the necessity is less, the wit is generally more dull.

"There is a feeling excited by the pure blasts that sweep over the mountains, which is utterly unknown to those who confine themselves to the plains that border their sides. This invigorating feeling is not to be described, and when the district which calls it forth is our birthplace, the affections are added to the delight which a stranger would experience, and the result is an elevation of mind which purifies the soul, and prepares us to die for our country should need be.

"In my walks over the Quantock Hills, my attention has been called to their products, their minerals, the woods which adorn their sides, the varied and magnificent views from their highest tops, the distant channel bounded by the Welsh mountains, and a thousand accessories which it is impossible to behold with common sensations; but far stronger are they when I reflect that I drew my first breath within their precincts, and that those from whom I derived my existence sleep beneath their soil. As life draws on, the memory of the past is added to the enjoyment of the present, and things in themselves common and trifling become sacred in our eyes.

"Over these wild and beautiful hills, at all hours of day and night, in all seasons, have I wandered, and never in vain. Mean must be the soul, and debased the heart of him, who is blind to the beauties of Nature, — the glowing language of the immortal God, the book which is spread wide to all; to which neither poet, painter, nor historian, can do justice; the emblem of that happiness to which we look forward in a future state.

"Independent of the beauties of Nature, there are many traces of generations long since passed away, particularly on Danesborough Hill, where are the remains of a fine Roman encampment. It has been said that Julius Cæsar ascended the highest point of the Quantocks, and that, on beholding the splendid view around, he exclaimed, 'Quantum ab hoc' ('How much is seen from this!'), whence the name Quantock. Whether or not Cæsar was really godfather to these hills, I must leave to the decision of learned antiquarians. The summits of the hills are too exposed and bleak of course to bear timber; but their sides are clothed with oak undercover, and in many parts with fine woods. Beeches, and, when they are planted, Spanish chesnut-trees, grow to a very large size, and occasionally very fine oaks are met with. In a tremendous storm which took place some time since, and which only lasted twenty minutes, I lost two beech-trees which stood side by side. One of them exceeded a hundred feet in height, and contained nine tons of timber. Thirty-five tons of beech timber were also blown down in its immediate vicinity. The spectacle during this short-lived hurricane was awful.

"The silver, Scotch, and larch firs grow to a very unusual bulk. Some years since, a group of silver firs, twelve in number, each averaging the height of one hundred feet, overshadowed my house. At the present time I have Spanish chesnut-trees which measure twenty-three feet in circumference. There is also a pollard ash now standing in Broomfield Green of similar dimensions. There is also an immense yew-tree in the churchyard, which is supposed to be a thousand years old; its circumference is upwards of twenty-five feet. I mention these facts to show that the Quantock Hills are very remarkably calculated for planting, if done judiciously. The ash flourishes so well in this district, that it is quite the weed of the country. No one who is interested in planting can observe, without pleasure, the tall, smooth, silver stems of this tree, rising up in parallel pillars, so as to form, in due time, a flourishing plantation. The great enemy of this tree in its youth is the rabbit, as the squirrel is the great destroyer of the larch and Scotch fir. It is much to be lamented that more attention is not devoted to planting. It is true that the time has been when the planter was far better remunerated than he can now expect to be; still, much may be done, and thousands of unproductive acres throughout Great Britain may be rendered highly advantageous to the owner and the public by enclosing and planting them.

"It has been a very false and exceedingly mischievous idea, to imagine that this island ever was, or is soon likely to be, over populated. As the planting of every hundred acres of otherwise unproductive soil, independent of the temporary employment necessary for the purpose required, is the forerunner of a great increase of labour; and every well-wisher of his fellow-creatures would rather see our hardy sons happy and prosperous at home, than wasting their strength in distant regions, incapable of assisting their country should an invasion by a hostile force take place. One of the principal mistakes of landed proprietors with respect to the culture of trees, is the suffering their young plantations to remain for years without being properly thinned out. Their gardeners and labourers hoe out the superfluous carrots and turnips, but little or no attention is paid to the thinning of plantations either young or old; indeed, they are in general irreparably injured before any mischief is dreamed of. Great care and incessant attention is requisite in this department, which will repay the planter to an extent little anticipated. One fine tree is worth a hundred small ones, both in value and beauty, and to produce this one many smaller must be sacrificed, to give the larger room to expand on all sides. It must be borne in mind that wood is mostly composed of carbon, and that the carbon of the tree is extracted principally from the atmosphere, by means of the absorbent power in its branches; and that if these branches have not sufficient room to expand, it is absolutely impossible for the tree to attain to any size. Again, it frequently happens that the cultivator, in order to preserve the number of his trees, lops off their branches unmercifully, and in so doing destroys the feeders of each tree, which these branches are. A vulgar notion prevails, that the branches to a great extent take off from the size of the body of the stem, so that one frequently observes tall thin skeletons of trees, denuded of most of their branches, disgracing the beauty of the landscape by their unsightly appearance. This is the case in many hedgerows, and is, perhaps, commonly done to prevent the fields from being too much shaded; but it is far better to dig up their trees by the roots, than to suffer such scarecrows to remain. Carbon is classed by electricians amongst the electro-positive substances, and in consequence it makes its appearance at the negative pole of the voltaic battery, as the opposite electricities attract each other. Now as the extremities of plants are in a negative state, electrically considered, they attract the carbon from the atmosphere, and thrive in proportion. A very simple experiment is sufficient to prove the rapidity with which carbon may be attracted from a substance yielding it, when exposed to electrical action. Form a voltaic battery of seventy cells, in the usual mode, and charge it with very dilute acid or water alone. Bring the extremities of two copper wires from each end of the battery, to the flame of a lighted candle, so as just to allow them to enter the flame, at a distance of about one-third of an inch from each other. In the course of a few minutes, the unconsumed carbon of the burning candle will be attracted to the negative copper wire and form a species of tree issuing from that wire, and within an hour or two will shoot out its black ramifications to the extent of one or two inches, or more, from the wire; while at the same time no arborisation will be formed on the positive wire. In this experiment the uncondensed carbon of the burning candle is projected mechanically into the atmosphere, and is *electro-mechanically* (if I may be allowed such a term) attracted to the negative wire; — whereas in the case of the withdrawal of the carbon from the atmosphere by the extremities of the leaves and branches of the tree, the carbon exists in the atmosphere united with oxygen in the state of carbonic acid gas, which passes chemically through the ramifications of the tree, the carbon being retained to feed the tree, and the oxygen liberated. This is an *electro-chemical* action, and not a mechanical deposit.

"As to the mineral formation of the Quantock Hills, they are principally composed of grawackè and clay-slate, in strata mostly dipping from north to south with transition limestone resting on their sides. The soil is more or less coloured by peroxide of iron. Metallic lodes traverse the hills, mostly in a direction from west to east. A mine was opened some years since at the western side of Stowey; it was called the Buckingham Mine, and was worked by three successive companies, and, as far as I can learn, several thousand pounds' worth of copper ore were at different times extracted from it. These ores, mineralogically speaking, were of very unusual beauty. The green carbonates of copper, or malachite, bore a high polish, and were only inferior to those found in Siberia. The blue carbonates were finely crystallised and of a beautiful colour. Both these kinds were found associated with large masses of compact and crystallised sulphate of baryta, stained with peroxide of iron. It must be noticed that, on some of the highest eminences of the Quantocks, similar tabular crystals of sulphate of baryta are found, but unstained with copper, which rarely makes its appearance at so great an elevation in this country; and a friend of mine found on Broomfield Hill, a specimen of that rare mineral, *cupreous sulphato tri-carbonate of lead.* A mining gentleman, of high reputation in Cornwall, paid me a visit of some duration, and in our walks over the Quantock Hills he gave it as his opinion, that at some future time this vicinity would become 'the principal mining district of the west of England.' * * * The real lover of science is rarely a covetous man, and not often a rich one. He reads the language of the Creator in all that surrounds him, and each successive discovery the more distinctly teaches him his immeasurable distance from the Throne of Truth: thus is his mind at once humbled and exalted.

"In my walks on the Quantocks I have often reflected how little we know of the habitudes of animals. How strong are their affections, how powerful their instincts — almost approaching to the confines of reason. When this great tyrant man is asleep, little dreaming of their gambols, then they quit their hiding-places, and revel in comparative security. Often have I stumbled on the red deer, while crossing the hills in the dead of night, or disturbed the fox with the light of my lantern. I never found any animal, except occasionally my own species, whom I could not tame by persevering kindness. Many strange adventures with bird and beast have occurred to me in these rambles, and once I nearly lost my life from a tremendous storm that overtook me at night; but I have been well repaid for all minor inconveniences by the opportunities afforded me of observing the phenomena of Nature. The startling meteor, the magnificent aurora borealis, the lunar rainbow spanning the horizon, with pale and mystical light, and, far above all, the refulgent planets rolling in their appointed course, and at a vastly greater distance, the ocean of starry worlds, whose size and numbers mock the telescopic calculations of erring philosophers.

"It is not merely in the enchanting month of May, when Nature calls for the song of the poet, or during the rich tints of autumnal scenery, when the purple heath covers the hills with its glorious dyes, and fills the morning air with fragrance; it is not merely at these times, that the range of the Quantocks demands admiration. Each season has its respective beauties and advantages, and even the sterile winter possesses adornments not to be found at other parts of the year. There are two distinct species of fog, which take place in the atmosphere. The one is comparatively rare, is highly electrical, and produces quite different effects from the other, or common fog, which contains so little electricity as only to be manifested by the aid of a delicate condenser. When the electric fog occurs in a freezing winterly night, each drop of moisture as it falls upon the meanest twig of a shrub or tree, or even on a blade of grass, in a freezing state, shoots out in a crystalline arrangement, forming a succession of radiations from the twig which becomes their centre; till at last the whole shrub or tree becomes covered with innumerable aggregations of acicular crystals of ice. I have seen a dead thistle supporting a crown of these needle crystals of at least ten inches in diameter. Now this is by no means the case, in an unelectrical fog, during which, in a night similar to the above (excepting that the electric agency is wanting), as the moisture is deposited upon the branches of the trees, upon each of which it is frozen by the severity of the cold instead of appearing in a crystalline form, as in the former instance, it simply cases the stem with a smooth cylinder of ice, like barley-sugar.

" I once endeavoured, however imperfectly, to describe in poetry the effects of this electric fog, in a magnificent sunrise which it was my good fortune to observe from the summit of the hills.

" 'Twas winter's depth, yet not the lightest breeze
Shook the keen icicle that gemmed the trees,
Which reared their stiffened heads in jewelled state;
Branches on branches, bowed with icy weight,
As drooped their lower limbs superbly bound
In radiant fetters to the spangled ground.
Feathered with heaven's own plumage, tipt with gold,
Glowing with dyes unnumbered, hues untold,
Stolen from the God of Day, who quits the hills,
And from his throne refulgent light distils.
Each tuft of thistle, in its gorgeous dress,
Scoffs at the laboured pomps that kings oppress;
From every centre emanating play
Its needled crystals in the blaze of day;

And each vile weed, which foot might trample down,
Laughs at man's art, and rears its starry crown.
Say, my bold Muse! canst thou presume to sing
Such highborn glories on thine earthly string?
Say, canst thou urge thine unremitting flight
From grosser darkness to the realms of light?
If so — tell how that red round orb unshrouds
His crimson face, and paints the blushing clouds,
Which ope their melting files, as he draws nigh,
While floods of ruby splendour drown the sky,
Save where some denser vapours stem his rays,
Empurpled islands in a sea of blaze.
Tell how life, light, and hope, and joy combine,
Sport in his beams, and from his chariot shine;
How the light fringe, which bound each stem in frost,
Flung back that fire in countless diamonds tost.
How the glad vales, their fleecy robe untied,
By slow degrees resumed their verdant pride.
No! the vast effort would untune my lyre,
Confound my song, and quench the rising fire!
Not our immortal bard, could he survey
The fleeting wonders of that blessed day,
The prostrate mind to such high theme might raise,
Or sweep his chords in all-sufficient praise."

To his friend, Mr. Kenyon, he thus writes: —

"Another year, my dear Kenyon, and another, and then — *sic transit*, &c., * * * * * * * * * Do you recollect the old lines you wrote at Seyer's. I sometimes, as if to punish myself, indulge in these painfully interesting recollections, and love, like the madman who scooped out his eye with a bone, to tear my heartstrings in keenly calling to mind days which never can return. The old court with its stone walls, the raised garden, the lilac bushes out of which I cut many a bow and arrow, the plain, neat, roomy house, the arch and study window, the row of elms and the two supereminent ones, which topped the whole, pointing out to many a schoolboy, returning from his vacation, that near that spot was to be his resting-place for the next five months — then almost loathed — now *consecrated;* yes, *consecrated* by the sad contrast of pains and disappointments, and by rising uppermost in every vicissitude of succeeding years: all this to look back upon, creates a feeling almost inexplicable — overpowering beyond description!" * * * * (Further on, in allusion to the death of a near relative of his friend, he says) "Alas! the hopes of man form the only reality in this world, and the enjoyment is removed from perspective, till death closes the scene. I have, nevertheless, a *firm belief* that 'whatever is, is best.' A Being who can make so beautiful a fabric as this world undoubtedly is, where all opposites are linked together so as to form one grand and harmonious whole, would never, in the plenitude of His boundless wisdom and benevolence, have raised up such a scourge as death, save but to answer the wisest of ends. It is to another existence we must have recourse, where, purified from baser passions, petty infirmities, we may humbly pray, and I trust not unreasonably hope, to renew our past friendships and affections, without dread of another cruel interruption. * * * * * When I was about four years old, I used to gaze upwards at the blue sky in a windy day, and long for a dispersion of what I deemed a misty blue, that I might peep into the glory of heaven beyond. When I was some ten years older, my fancy placed heaven first in the sun, then in some remoter star, then in the space beyond, then beyond human sight — now it seems to me that heaven consists of a boundless sight and full comprehension of all the mighty mysteries which the Almighty has, in His infinite wisdom, cast around us, and that could we be permitted to comprehend all that is above, and below, and around, then indeed should we find ourselves in a heaven of heavens; but it strikes me that comparing my former contracted ideas with my present more expanded ones, that by pushing on in like manner, my present conception of the glories which exist, must be mean and contemptible in comparison to the *vast reality*, and I shrink within myself at my own insignificance." *

* * * * * * * * *

"Believe me, dear Kenyon,
"Yours very sincerely,
"ANDREW CROSSE."

Mr. Crosse was at a very early period an advocate for educating the humbler classes, and of disseminating knowledge generally. He used to say, "There is no real difference between men but education." Some of his written remarks on these subjects bear date about this time. The following extracts are made as expressive of his opinions.

"The life of man is so short, and short as it is, so distracted by unavoidable occupations, cares, sicknesses, and temptations, to say nothing of the time necessary to rest his mental and bodily powers, that one possessed of every qualification requisite for the most speedy and effectual attainment of knowledge, has, from a host of hindrances arrayed against him, the mortification of seeing his precious time consumed ere he has advanced a small way towards the goal, and of finding himself vibrating on the brink of the grave in the midst of a thousand vast plans and resolutions. 'Thus far shalt thou go, and no farther!' seems to be the language held out to him by the Creator. A prospect of indescribable magnificence invites him. Death and corruption bar his path, at least in this world. It is only by slow degrees, by first searching for a foundation whereupon to plant the ladder of science, by making sure of one step before he advances to a second, by carefully examining the way as he proceeds, by the most determined resolution, patience, and perseverance, that he can hope to make any considerable progress. Yet, on the other hand, comparatively speaking, much has been done, and much may be done, by a single individual in whom the necessary qualifications are united. It should be superfluous to notice such great names as those of Newton and Davy, and others who have been an honour to the age in which they flourished. It may be right however to remark that there are few but may contribute their mite to the cause, and then advance a considerable way in it, if they possess sufficient time, ardour, and perseverance. Talents are not so rare, one man is not so raised above his fellows in intellectual capacity, as to cause the mass of mankind to despair of attaining a considerable degree of knowledge. * * * * * Well, then, be it granted that knowledge is desirable. Is there one man, however poor, however abject, crawling on the face of God's earth, from whom we ought seriously to shut it out? to keep him in ignorance of the glories by which he is surrounded, and of the capacity he has for enlarging them? to prevent him from digging for that treasure which he has a right to possess in common with his kind? to chain his intellect, but set free his limbs? to allow a being endowed with superior intelligence to veil his best faculties in sleep, or, if awakened, to be used only for the furtherance of mischievous propensities. Reason scarcely awake is a curse rather than a blessing. It mistakes cunning for wisdom, subtlety for sound argument, unjust policy for true justice. * * * It is the aim of those who aim at true knowledge to have no concealments, to labour sedulously and cheerfully at their work, and to invite all without distinction to unite in the accomplishment of the task. Not to endeavour to mount the summit of the hill for the paltry pleasure of laughing at the scramblers below, but to love wisdom for the sake of its Contriver, for the sake of itself, for the sake of imparting it to others. Nor is the acquisition of knowledge likely to puff up the mind with ridiculous ideas of our own importance; each fact gained will tell us how many more are to be learned. If we have ample time to dedicate to study, it will thus be usefully and profitably employed; if we can only steal a portion of each day from our necessary avocations, we shall enjoy it with redoubled delight, much as a townsman feels a greater zest in a walk through the country than one to whom such scenes are familiar. Still, the pleasure to be derived from those pursuits differs from other pleasures: there is no satiety, no palling upon the senses. Happiness is to be sought *internally*, not *externally*: and I appeal to those who have devoted themselves to science or to literature, if they have not found in those powerful but fascinating pursuits a charm far greater than any derived from an occasional indulgence in other pleasures; a delightful, but never-failing consolation to which one flies for refuge from the hypocrisy and ingratitude of false friends, and the annoyances inseparable from our condition on earth. The mode of acquiring knowledge, as well as the knowledge to be acquired, and the means of disseminating it, is therefore of permanent importance to the interests of man: first, that he deceive not himself; next, that he deceive not others; finally, that he use every exertion in communicating to his fellow-creatures those advantages which he feels to be of such infinite service to himself."

The following poem on "Poland" received the warm approval of the poet Campbell; two or three others which I have ventured to add were also written about this date.

"POLAND.

"THERE is a name—'t is written in His breast
Whose eye ne'er closes in forgetful rest;
'T is known in Hell, where shrieking despots feel
Its torturing echoes, keen as barbs of steel;
'Tis seen in History's remorseless page,
At once the blot, the glory of the age;
The winds indignant sweep it o'er the main,
The mournful shores return the sound again;
Minstrels unborn shall fit it to their lyre,
And future nations thrill with kindred fire;
Banners shall bear it to admiring eyes,
And music waft its chorus to the skies.
'T is freedom's watchword, and shall millions wake
A just and fierce, though slow revenge to take;
'T will loosen diadems, and scatter thrones,
And for plebeian tears draw royal groans.
Poland, 't is thine, the unconquered still in mind,
Though in corporeal slavery confined!
Though Russia's demon would in pride elate
By edict stern that name annihilate!
Though kings in monstrous policy unite,
And spurn at treaties to destroy thy right;
Kings who, from plunder, holy temples rear,
And while they worship, shake th' ensanguined spear.
Yes! let them swell the reverential song,
And through polluted domes the strain prolong;

Themselves exalt, the righteous and the wise,
Faultless as gods, the world their sacrifice:
Fear they not vengeance, when the tempest calls,
And Heaven's resistless bolt in judgment falls;
When nations rise, and one united voice
Of *death to tyrants* bids the earth rejoice?
Yet the dread time shall come, as sure as Fate,
And despots crouch when crouching is too late;
When in the monarch man may spy the knave,
And ask what spell so long confined him slave.
Say, Poland, where are now thy glories fled?
Will the deaf soil give back thy martyr'd dead?
Will the stern Czar his bloody deeds deplore,
And from Siberia's wastes thy sons restore?
No! victim to unmitigated woes,
Clutch'd in the grasp of unrelenting foes,
Unlike Judea's captivated lyre,
No song from thee thy conquerors require;
But bolts and barbarous chains thy strength control,
And deep the iron vibrates in thy soul!
Is 't not enough for wretched man to face
The natural ills inflicted on his race,
In joy to tremble, or in torture smart,
Till certain death receives his bursting heart;
But that his brother should, for fancied gain,
Heap load on load, and agony on pain!
That his own flesh and bone, when robed in state,
Should plan new sufferings, new woes create!
Tread on the trodden, triumph o'er the weak
And on the already punish'd vengeance wreak?

Yet such is justice, when the one decrees,
And one frail nod bows down a thousand knees.
Can Amaranth hues or flashing gems repay
For realms undone, and towns to flame a prey?
Can all that woos the enamour'd sense to sin
Be worth the maddening horror it may win?
Or does it swell with pride a tyrant's veins
To hold the noblest race in hopeless chains?
If so, then well may Poland's scourge employ
His countless hosts their victim to destroy!
Relentless fiend, who spurnest every tie;
Deaf to the mother's wail, the orphan's cry;
Sear'd to the bride's lament, the veteran's tear;
Blind to the cheek of beauty pale with fear;
Dead to the warrior's scowl, the widow's prayer,
The father's contumely, the child's despair;
Compared to thee was Nero truly great —
He fired a city, thou expung'st a state!
Sleep Europe, sleep! Gaul, gentle Britain, sleep!
Let the soft torpor through your vitals creep!
Sleep on in dreamless unison, nor heed,
While sweet your rest, what distant cities bleed.
Sleep on, or, should ye dream, in dreams behold
Not the earth's God, but yours — enchanting gold.
Let worldly calculations guide the rein,
And in due bounds the indignant soul contain!
Thus may your prudence, fears, and wit combine
To save the farthing, but expend the mine.
Sleep on! till northern thunders wake your shores
And Retribution's final cannon roars;

Till justice, close at hand, in funeral tone
Asks, why so long unheard was Poland's groan?
Then may you cry, bereft of power and pride,
For that same aid, to others once denied!
Perchance the tender Muscovite may hear,
And smiling Cossack check his swift career!
Ah! could the bard's aspiring strains avail,
The world would shake, convulsed at Poland's tale,
Earth's sceptred plunderers reel in wild alarms,
And nations spurn their bonds, and shout, To arms!
To arms! to arms! from hill and town and plain,
From the rude mountain and unshackled main;
From cliff where nature reigns in savage state,
Where whirlwinds blow but to invigorate;
From tropic glade by spicy zephyrs fann'd,
From snow-clad pinnacle, from scorching sand;
From vale, whose lilies shame the pomp of king,
From grove, whose songsters spread their free-born
wings;
From isle unbounded but by sky and wave;
From freedom's sail, from Kosciusko's grave;
To arms! to arms! should peal o'er land and sea,
Nor cease the sacred sound till Poland should be free."

"THE TYRANT AND SLAVE HOUR-GLASS.

"Come! set the time-glass on the stand,
And let us mark the lessening sand!

Tyrant at top, and slave below,
An hour will change the passing show.
Invert it, and in order due
The slave becomes a tyrant true,
And he that took the sovereign lead
Now bows in turn his abject head;
Each fitted for the double part,
A moving thing without a heart."

"TO THE ARISTOCRAT.

"Whence is thy charter, man of power!
That hosts must crouch below thee,
Must heed thee as a moated tower,
And at a distance know thee?

"That thou in thy vast brain must mould
The million at thy pleasure;
That *mind* must bow before thy gold,
And matter be our measure.

"Think'st thou thy sight alone can reach
The stars that glow around us?
Or that thy wiser tongue can teach
That which might else confound us?

"Is it that all save thee are stone,
When music bursts her slumbers,
That pleasure wakes for thee alone,
And pain for countless numbers?

"That the submissive orb of day
Pours forth for thee its lustre!
For thee the ocean billows play
For thee the fruits rich cluster!

"That for thy smile all nature waits,
As fearful of undoing!
That at thy frown the vengeful fates
Must wrap the globe in ruin!

"That the ignoble mass are tools
Framed for thy scorn or laughter!
That hell is for plebeian fools,
And heaven for *thy* hereafter!

"Think'st thou? but no, thou dost not think,
No thought can near thee venture!
Thy venal lust has closed each chink
Through which the light may enter!

"Then round thee draw the mystic line,
The fog of rank be o'er thee!
The powers of darkness, be they thine!
Long may their sons adore thee!"

"HUMILITY AND DEFIANCE.

"WHAT thoughts conflicting in my bosom rise!
This strikes me down, *that* lifts me to the skies!
Now I recline an infant at the breast,
Now stride a warrior with forbidding crest.

"Here grovel low a helpless, earthly clod,
There pant defiance to the oppressor's rod;
At first with not a finger to oppose,
Then every pulse with hostile fury glows.

"Or, soft as rills which pour their sacred stream
In nightly murmurs on the poet's dream;
Or firm as rocks whose echo laughs to scorn
The puny summons of the huntsman's horn;

"The windows of my soul at once reveal
A twig of osier and a bar of steel:
Thus good and ill, and light and shade, combine,
And, though distinct, in folds together twine.

"But lives there one, in solitude or throng,
So versed in practice, or in wit so strong,
Whose eye scarce human clearly can decide
The secret links which right and wrong divide?

"Who by some mental microscope can show
Where virtue ends and vice begins to grow?
Can dive into the mazes of the mind,
All doubt annihilate, unfilm the blind?

"Point out so far, and not beyond to steer,
Where to press boldly on, and where to fear?
Where to submit and due allegiance pay,
Where to resist the ruthless spoiler's sway?

"Where to be hot as fire, or cold as ice,
Where life is infamy, where death is vice?
If such there live, in planet or in fame,
No mortal lineage can his kindred claim.

"He, only He, our inmost thoughts can tell
Who rules alike o'er Heaven and Earth and Hell.
To Him I bow, before his awful shrine
Each favour'd wish, each rebel thought resign.

"If He but wills, and I that will can see,
That will be life and death and all to me!
But not o'er me shall man tyrannic reign,
I scorn his bondage, and I rend his chain!

"'T is true my viler limbs he may control;
These be his share, but God's and mine my soul;
My soul which, in its deep recesses, hates
Spoil-nurtur'd conquerors and crouching states."

CHAP. III.

RELATIONS WITH THE WORLD.

1821—1846.

POLITICAL matters engaged Mr. Crosse's attention a good deal, about this time. I find a MS. of a paper headed, "Ought an excess of agricultural produce to be the cause of a proportionate degree of poverty to the cultivators, and consequently the owners of the soil?" In this article, he strongly combats the idea that excessive produce, or increased population, is an evil. He always considered that legislative emigration was the greatest possible mistake, and that it was the way to weaken the country, and destroy the very sinews of her existence, by sending away her able-bodied labourers, who, when necessary, recruit the number of national defenders. This opinion he never altered.

The agricultural distress of this period was deeply felt in Somersetshire. In January of the year 1822, there is convened at Taunton an agricultural meeting, numerously attended by the "gentry, magistrates, and freeholders." A petition to Parliament to the following effect is proposed. "To the Right Honourable the Lords Spiritual and Temporal, in Parliament assembled. We, the gentry, clergy, freeholders, and occupiers of land in the district of the once opulent vale of Taunton, most humbly represent to your honourable House, that the cruel distress throughout the district in which we reside has arrived at an unparalleled height, and is daily increasing to an alarming extent. That the progressive decline in the value of all productions of the earth, accompanied by an overwhelming burden of taxation, such as never was endured by any country, has swallowed up the capital of the farmer, and brought the greater proportion of independent yeomen to the brink of ruin, which, without the most speedy relief, must terminate in the annihilation of that most excellent and invaluable body of men. That your petitioners therefore pray, that the strictest economy and retrenchment may be enforced in every department of the State." This petition is seconded, but the speaker in doing so does not give it his unqualified approval; he thinks that these "grievances should be explained." "An oppressive taxation is the evil under which the whole country is suffering, and it is in vain to expect that war taxes and peace prices can exist together." A clergyman comes forward, and says that in his parish "the rates absorb the reasonable gains of industry: in 1793 the poor-rates were 821*l.*, in 1821 they amounted to 2000*l.*" Opinions differ, the reverend gentleman considers that a prominent source of the present evil is "the inadequate duty on foreign corn." Another member of the same cloth actually moves a resolution "that all restriction whatever on foreign corn be taken off," as a first step towards laying a firm and stable foundation for the agriculturist, on which to rest his future hopes. But the resolution, after much clamour of disapprobation and applause, was negatived by a large majority. The time was not come.

The same reverend gentleman told the meeting that the petition would be of no avail, "and would share the fate of thousands of other petitions, which were like so many murdered babes smothered in parchment. The people of England were wearing fetters which they might break, if they had but the spirit to make the attempt. The reformation of the House of Commons could alone afford any hope to the country, of an improved condition."

The war has ceased seven years, the benefits of peace have not accrued, the arts of peace have not flourished; "500,000 of the most useful class in the country must be ruined," says a land surveyor: as a farmer, his stock "has declined in value, in two years, from 4000*l.* to less than 3000*l.*"

Mr. A. Sandford, the chairman of the meeting, feelingly says, "Would to God that those who deny the existence of the miseries of the agriculturists, or doubt their unhappy extent, would enter the habitations of the yeomen of this once opulent county, and witness the privations to which they are now reduced; their hopeless days and sleepless nights; their sorrowing families; and their exhausted means of bestowing on the objects of their affection the comforts to which their honest industry had entitled them."

Such is the sad state of things which all agree in wishing to have altered, but upon the ways and means to effect this change they do not agree. One gentleman mentions this startling fact. In 1792 prices were much about the same as they are now (1822), the taxes were 16,000,000*l.*, the taxes of 1820 were 70,000,000*l.*

Mr. Crosse addresses the meeting in these words: "Though not in the habit of public speaking, I should deem myself a traitor to my country if, at an awful crisis like the present, I was to refrain from expressing my sentiments. I see infallible ruin before me. I am perfectly free from all political opinions, and speak the sentiments of an unprejudiced man. I am in the habit of intimacy with many distinguished persons, both Whigs and Tories, but am free to think and act for myself; and my opinion is most decisive, that absolute and irretrievable ruin must overwhelm us unless we are saved by an efficient and constitutional Parliamentary reform." Mr. Crosse then expressed his admiration of, and attachment to, the British Constitution, and pronounced a warm panegyric on the clergy of the Established Church. He continues: "I have no malevolence towards ministers, but I am perfectly convinced that all the miseries with which the country is inundated are solely to be ascribed to the Pitt system, which has been wrong and pernicious from beginning to end. So long, therefore, as the present ministers remain in power, real improvement in public affairs cannot be expected. Within a few years the taxes had been quadrupled, and three-fourths of them must be taken off before their pressure would cease to be intolerable. One half of the National Debt must be reduced, for, having been created by a fictitious currency, it is impossible to cope with it now we have returned to cash payments: had I been more accustomed to public speaking I should have given my feelings a better utterance, but I have honestly and conscientiously avowed my opinions." He then, with much feeling, benevolently exhorted landlords at this period of distress to be merciful to their tenants, and tenants to be industrious to their landlords. "If members of Parliament and ministers will show no example of economy, it behoves every individual to set them an example of retrenching any superfluous expense in private life."

His warm feelings of consideration for others, the utter absence of all selfish prejudices of class, together with an unbending integrity, made Mr. Crosse a great favourite with the people generally. "You have immense influence in the county," said a brother magistrate to him one day. "I despise all influence," replied Mr. Crosse. "I don't think I possess any; and if I did, I would not exert it; I only desire the good of my fellow-creatures."

Those who remember this period will not soon forget the stormy discussions, the bitter animosities, and the strong tide of opposition which preceded the passing of the Reform Bill: all this is now matter of history, but events of the day have their effect upon characters. Mr. Crosse was generally thought to be an ultra-liberal, but I do not consider that, strictly speaking, he was attached to any *party;* his opinions partook somewhat of the republican views of his father, still more of his own poetical nature, which, together with a general philanthropy, rendered him a character fitted to inspire and elevate the people in a time of political excitement and social depression: he was an honest man, and a man of noble and generous sentiments; and, bad as satirists tell us the world is, it always responds to what is pure and good: there is something in human nature, even the human nature of venal boroughs, which awards an almost involuntary acknowledgment to honest worth. "The character of the speaker," says Aristotle, "affects the audience."

Mr. Crosse continued to take a very active part in the politics of the day. On several occasions, he proposed or seconded the candidate for the county; he was long the political ally of his friend and neighbour, Mr. Tynte, of Halswell Park; but so little did party feelings influence him, that once at a Whig dinner, he chanced to be sitting next the president, and he proposed, in an undertone, to him, that they should take this public opportunity of drinking the health of a political adversary, who had acted very nobly in some public manner; but the reply he received was, "Better not, better not; it won't do." "I felt," said Mr. Crosse, "that I was not understood,—I believe I was silent for the rest of the dinner; how could I sympathise with such ungenerous prejudices, even though the opinions of these men were the same as my own?"

On one occasion, he had been speaking very boldly on the hustings, when some one from the opposition called out, in derision, "Oliver Cromwell! Oliver Cromwell!" "Gentlemen, I thank you for the compliment," he quickly replied; "and if I were Oliver Cromwell, I would sweep all such as you from the face of the earth," and, suiting the action to the word, there was a simultaneous though but momentary movement on the part of the swaying crowd beneath him. When the meeting was over, a gentleman, covered with blue ribbons, came up and shook him heartily by the hand, saying, "Why, Crosse, you don't care for the whole world." "Not if I think they're wrong, and I am right," answered the enthusiastic speaker. So little animosity was personally felt to him, that, I think on this very occasion, the platform being quite full, and having literally only room for one leg to stand upon, his other leg was supported, and his equilibrium preserved, by a "red-hot Tory," as he designated him: his kind supporter was not so fond of a practical joke as himself, I imagine, or he could scarcely have resisted the temptation of putting a ludicrous ending to the fulminations that were going against his own party. Some persons in a different grade of society were, however, not quite so liberal; for when he attempted to speak, a knot of farmers hissed and hooted, and would not for a long time suffer him to be heard. A stranger, a commercial man from the north of England, noticed the extreme wrath of this particular group, and turning to some one near him, said, "Why are these farmers so angry with that gentleman? who is he?" "Why, don't you know him? that's Crosse of Broomfield,—the thunder and lightning man; you can't go near his cursed house at night without danger of your life; them as have been there have seen devils, all surrounded by lightning, dancing on the wires that he has put up round his grounds."

A gentleman lately mentioned to me an expression of Mr. Crosse's, at an election, the recollection of which, though so many years had passed over, had not faded from his memory, so deep was the impression it made,—the words were simply these: "I value one ounce of knowledge more than a ton of gold; but I value one grain weight of human kindness more than a ton of knowledge." "This sentiment," said my informant, "struck me particularly, and the more so as it was in the midst of a contested election, when all the most angry passions of our nature are struggling together."

The following fragment, called An Account of the Island of Elattosis, is an endeavour to work out a curious idea which Mr. Crosse had of representing a place where death was unknown; and where, instead of dying, people and things should diminish; but he broke off in the midst of his narrative. "For," said he, "I found out how impossible it is to represent anything contrary to nature."

"THE ISLAND OF ELATTOSIS.

"As I was reclining comfortably in an easy elbow-chair in my study, in the remote part of a lone country house, a cheerful wood fire blazing brightly before me, and the candles standing unlighted on a small reading table, though the increasing intensity of the evening shadows had long reminded me of my neglect, I could not be but struck at the contrast of the comforts which surrounded me, with the rude howling of the blast, as it swelled or died away amongst the shapeless masses of wood which environed the mansion, and which were scarcely discernible through the window, not merely on account of the approach of night, but from the hasty pattering of large drops of rain against the casement, which the violence of the gale alone prevented from descending in a settled shower. How many poor wretches thought I, are at this instant exposed to the fury of the tempest: houseless, penniless, without food, or sufficient to cover them, sinking under every sort of distress and disease. Perhaps a ship just returning from a long voyage, laden with the wealth of the Indies, and filled with passengers ardently desirous to welcome their native land, now strikes upon a hidden rock, and, in lieu of an endearing and hospitable reception, is ushered into horrors, whirlpools, and destruction,—whilst I, quietly seated in my chair, stretch my legs at my ease, and—stop, cries Recollection,—Death will soon stretch your legs, and whether one is suffocated in a muddy ditch, drowned in the ocean, or pines away upon a feather bed, with half a score of sage-looking doctors, mumbling politics in a corner, still Death is Death, and bow to it we must. Dreadful and tremendous decree, thought I, could not some other less agonising method of causing the disappearance of the human race have been determined? Scarce had the impious thought suggested itself, when a thick mist instantly filled the room; the howling of the wind subsided, the rain ceased to patter, the fire in the grate no longer crackled, nor did the flame move. All was breathlessly still, and a horrible expectation came over me of something unutterable.

"On casting my eyes towards the lower part of the room, I clearly discovered the head of a venerable old man, whose white beard seemed to unite with the thick mist which pervaded the apartment, but which had not the power of concealing from my view his majestic though stern countenance. 'Worm,' cried he, 'canst thou pretend to know the events which take place in other regions? In the remote Isle of Elattosis, death is unheard of, and yet terror and anxiety are no more strangers to its inhabitants than they are to those of your regions.'

"'How can this be?' I ventured humbly to breathe.

"'See,' cried he; 'know by what no one can bequeath —experience.'

"Immediately my sight failed me, and methought I sank into an ocean of waters, which seemed to bubble round my ears. In a few seconds, as I imagined, I found myself standing alone on the sandy shore of an apparently fertile country, whose swelling hills, variegated with wood, meadow, and corn land, rose one behind the other, till the whole was blended in a dark blue horizon. I had but little time for reflection, when a man dressed somewhat like an English mariner came up to me, seeing I was a stranger, and saluted me in courteous terms, to my utter amazement, in the ancient Greek tongue. As I had had some knowledge of this language flogged into me when a boy, although I had not much idea of chattering it, I was enabled, after a little consideration, to enter into some sort of conversation with him.

"'Pray,' said I, rather abruptly, for I had not learned to be very civil during my sojourn in England, 'what the devil is the name of this queer-looking country?'

"'Friend,' replied he, calmly, 'it is the Island of Elattosis, and you are the only individual whom I ever met with who was ignorant of it. I suppose you have sailed from some of the distant isles which lie at the north-west of this land?'

"To this I assented with a nod, as I thought it the wisest course to pursue, and I was used to humbug in my own country, and made it a rule, as seldom as possible, to appear ignorant of anything.

"'Sir,' said he, 'I am a captain of a small merchantman, and am well acquainted with these north-west islands, the Eikosi, the dress of whose inhabitants, I must confess, does not much resemble yours; and yet I know not from what other country you can have arrived.'

"'Captain,' returned I, with a bold face, 'I allow I am clothed in a strange dress, but it is the consequence of a wager, which I shall lose if I adopt any other during the next three days, at the end of which period I intend to resume the dress of my own country.' I could not avoid reflecting on the inconvenience of one falsehood, which becomes necessarily the parent of a host of others.

"The Captain, observing me look serious, said with much kindness, 'Come, sir, as you are at all events a stranger in this land, and if you can put up with sailors' fare, you had better accompany me home.' To this I gladly assented, with thanks for his hospitality; and we set off on our walk. As we proceeded, after crossing a few fields, we came to an open road, bounded on one side by a down or common, entirely covered by what I imagined to be a species of moss. On stooping down to examine it, I was surprised to find it consisted of a kind of moss, each stem of which exactly resembled an oak-tree in miniature, with its leaves and branches complete. Accordingly, I knelt down, and applied a pair of pocket scissors to half-a-dozen stems, intending to cut them off.

"'Stop,' cries the Captain, 'what are you about to do? it is not the custom here, for strangers to destroy the timber on a gentleman's estate.'

"'Timber!' cried I; 'surely, you don't call this brushy moss, timber?'

"The Captain shook his head, and made no answer, but looked as though he suspected I was not in my right senses. Finding that something was wrong, I pocketed my scissors, and we continued our walk, till he came to a dwelling-house of moderate size; this, my companion informed me, was his home. 'My family,' said he, 'is but small. It consists of an old father, a sick wife, and two children. My wife is so exceedingly ill, that I am obliged to make use of a magnifying power of 17 to discover her; but the doctor gives me great hopes that she will be restored to her original size. As for my poor old father, he is so reduced by age and ill health, that he does not exceed fifteen inches in height at present.'

"Here my astonishment was extreme, and I began to fear that I had gotten into the company of a maniac. On our arrival at the Captain's house, we were met by two chubby-faced children, who stared at me with great surprise.

"'How is your mother, my dear?' said the captain, to the eldest.

"'Oh, father, she is much better. I have looked in the glass in which she is kept, and she is grown to the size of a broad bean.'

"'Heaven be praised!' said the worthy Captain, 'bring her to me, John, carefully.' John was in an instant out of sight, and soon returned with a half-pint tumbler in his hand, in which the Captain desired me to look. No language can describe the surprise I felt on discovering a minute female figure, dressed in a loose robe, and not exceeding an inch in height. This little fairy form was walking rapidly round the bottom of the glass.

"'Ah, poor thing!' said the Captain, 'she is not yet large enough to make her wants known by her voice. Put her safely on the mantelpiece, my boy, and tell the nurse to drop into the glass some grains of boiled rice. But I beg pardon, sir,' said my host, 'for my apparent neglect. Our dinner must by this time be ready; pray walk this way;' and I followed him into a plain but neatly furnished apartment, where we partook of a boiled leg of mutton, and some mashed turnips and potatoes."

* * * * * *

"I had proceeded so far in my story," said Mr. Crosse, "when I dashed down my pen, conscious of the absurdity of attempting to describe any state of things so unnatural, that death and destruction should not be known; in truth, I could not, for I had myself violated the plan, and allowed the existence of death, in the fact of the food they eat; for, even if I had made them out to be vegetarians, still there would have been destruction going on; and so indeed would it be, should I but suffer them to breathe the very air, filled as it is with invisible life. This rude attempt of mine taught me one lesson — that is to say, how impossible it is for us to realise anything out of our limited experience of conditional truth."

I shall never forget a striking instance of the power of mind over body which occurred to Mr. Crosse, who, I should observe, was in general a singularly nervous person, and indeed, from the age of fifteen to the time of his death, he suffered more or less from nervous attacks, which were so distressing that sometimes for half an hour he would experience all the agonies, and worse than the agonies, of dissolution; for his powers had of course that vitality which makes all sensation trebly acute. He said, in one of his letters to a friend, "The ill health I suffered in my younger days made me imagine that I should never see thirty." One of these terrible attacks he thus describes in poetry, his most natural mode of expressing all strong feelings.

"ON A SUDDEN ILLNESS.

"So, Death, by my frail door thou fleetest,
And, as thou passest by,
With warning hand thou lightly beatest,
To hint that I must die!

"And though thou say'st not at the portal
When thou shalt step inside,
'T is kind to tell me I am mortal,
And check me in my pride!

"How many a tree my fathers planted,
I' ve lived to see decay!
How many a flower whose scent enchanted,
Is past for aye away!

"How many a string whose tones enthralled me
Is rudely rent in twain!
How many a voice which sweetly called me
I ne'er shall hear again!

'And while creation dies around me,
Shall I escape the tomb,
To herd with those whose sufferings wound me,
Yet flee the general doom?

"Have I not closed eyes which were lighted
With joy as I drew near?
Have I not seen those prospects blighted
On which the sun shone clear?

"Have not I felt, since Nature made me,
Far more than death can bring?
For should a thousand deaths invade me,
The past would blunt their sting!

"Say, does not all that floats about me
Scoff at life's vain pretence?
The worm I trample seems to flout me,
And bids me to go hence!

"The cloud above, the wave below me,
All beckon me away!
Each sight I gaze on does but show me
That here I must not stay!

"Better to writhe in death's last anguish,
With hope of some relief,
Than in this weary world to languish
The sport of joy and grief!"

Mr. Kenyon used to say, "Crosse, I don't like your verses; they tear one's heartstrings." But the anecdote I am about to relate shows the power which reason had over even this nervous susceptibility of temperament. Mr. Crosse was returning home one day by the side of one of the ponds in the grounds of Fyne Court, when he saw a cat sitting by the water. In the spirit of boyish mischief, which never forsook him, he sprang forward to catch the animal, with the intention of throwing her into the water, but, to use his own words, "She was too quick for me to catch her, but not quick enough to escape me altogether. I held her for an instant, and she turned and bit me severely on the hand. I threw her from me, and in doing so, I saw that her hair was stivered; the cat was evidently ill. She died the same day of hydrophobia! The circumstance passed from my memory as weeks rolled on; but about three months afterwards, I felt one morning a great pain in my arm; at the same time feeling exceedingly thirsty, I called for a glass of water: at the instant that I was about to raise the tumbler to my lips, a strong spasm shot across my throat; immediately the terrible conviction came to my mind that I was about to fall a victim to hydrophobia, the consequence of the bite that I had received from the cat. The agony of mind I endured for one hour is indescribable: the contemplation of such a horrible death — death from hydrophobia — was almost insupportable; the torments of hell itself could not have surpassed what I suffered. The pain, which had first commenced in my hand, passed up to the elbow, and from thence to the shoulder, threatening to extend. I felt all human aid was useless, and I believed that I must die. At length, I began to reflect upon my condition. I said to myself, either I shall die, or I shall not; if I do, it will only be a similar fate which many have suffered, and many more must suffer, and I must bear it like a man: if, on the other hand, there is any hope of my life, my only chance is in summoning my utmost resolution, defying the attack, and exerting every effort of my mind. Accordingly, feeling that physical as well as mental exertion was necessary, I took my gun, shouldered it, and went out for the purpose of shooting, my arm aching the while intolerably. I met with no sport, but I walked the whole afternoon, exerting, at every step I went, a strong mental effort against the disease: when I returned to the house, I was decidedly better; I was able to eat some dinner, and drank water as usual. The next morning the aching pain had gone down to my elbow, the following it went down to the wrist, and the third day left me altogether. I mentioned the circumstance to Dr. Kinglake, and he said, he certainly considered that I had had an attack of hydrophobia, which would possibly have proved fatal had I not struggled against it by a strong effort of mind."

"My dear Kenyon,

* * * * * * * * *

* * "On looking around me, I cannot but be struck with the apparent little value which the Creator seems to set upon human life. We are mowed down (to say nothing of other countless worlds) by myriads, heaps on heaps, by famine, sword, pestilence, and tortures of all kinds, mental and bodily. Wit, stupidity, gentleness, brutality, science, ignorance, all jumbled together, tossed on a common dunghill; but out of which shall rise, if God please, all that is valuable and pure, clothed probably with senses of which we have at present no clearer idea than the brute has of the thoughts of his master, but fitted to grasp with one view the panorama of magnificent perfection which may surround us, quitting for ever the gross, unintellectual part of our natures, all that in this detestably selfish world teaches man to look down upon his fellow-creatures as a reptile only fit to prey upon; all that prevents him from making a small share of that allowance for his neighbour's follies which each of us will groan for at the bar of Eternal Justice. But a truce to sermonising: you and I are at a time of life when it is excusable in us to take a little time to fold our garments decently around us, ere we fall as Cæsar. By the bye, Kenyon, I walked to Mullins',

a fortnight since, within thirty-eight minutes, being four minutes less than I have ever done it before. Mullins calls the distance four miles, so that you see I am not yet legless. I am now become a regular, or rather irregular, pedestrian, not having had possession of horse or mare during the last two years. I find this plan agree with my health and pocket, the latter of which is somewhat necessary to attend to, as my two elder sons are at their respective colleges. * * * My brother is metaphysicising at his cottage two miles from me. * * * The day before yesterday I passed with Tom Poole, who desired to be kindly remembered to you. I am glad to hear that your brother is become an ultra-Liberal. My brother and I are determined Reformers, but have little time to waste on politics. I have lately constructed a voltaic battery of ONE THOUSAND AND TWENTY-FIVE pairs of metallic plates; also an electrical battery, composed of talc plates coated with tin-foil. This last battery being interposed between the poles of the voltaic battery, charged with common pump-water becomes *instantly* charged, and to an intensity sufficient to deflagrate metallic leaves, explode fulminating powders, cause iron wire to perpetually scintillate, &c. &c. I have hopes to be enabled thus to form an apparatus capable of giving perpetual LIGHT, HEAT, and MOTION. I have likewise made some very interesting discoveries in electrical crystallisation, having produced cubes of metallic silver, and four-sided prisms capped with four-sided pyramids of muriate of mercury, from their respective solutions, by means of slow electric action. As for my poetry, I do no not attempt much refinement; the age is too refined already; and I feel a sort of loose carelessness and sturdy defiance, which but ill amalgamate with the courtier-like pliability and aristocratic exclusiveness of the times. My days are fully occupied in poetry, politics, natural philosophy, planting and pruning, and a thousand *et ceteras.* I have just given two lectures on atmospheric electricity to the Taunton mechanics, and have just discovered a method of doubling the power of cylinder electrical machines, both cheap and simple. * * Charles Tynte has invited me to make a tour through Ireland with him, and afterwards to Switzerland and Italy, but I am a fixture—erratic in mind but chained in body.

"Believe me, dear Kenyon,
"With every kind wish,
"Yours ever sincerely,
"ANDREW CROSSE."

In the year 1828, I find, by memoranda, that Mr. Crosse thought and experimented a great deal on atmospheric electricity. The following theory of the construction of a thunder-cloud, I subjoin in the same words as he himself detailed it, in an account furnished for publication in a late work on electricity.

"On the approach of a thunder-cloud," says Mr. Crosse, "near the insulated atmospheric wire, the conductor attached to it, which is screwed into a table in my electrical room, gives corresponding signs of electrical action. In fair cloudy weather the atmospheric electricity is invariably positive, increasing in intensity at sunrise and sunset, and diminishing at midday and midnight, varying as the evaporation of the moisture in the air: but when the thunder-cloud (which appears to be formed by an unusually powerful evaporation, arising either from a scorching sun succeeding much wet, or *vice versâ*) draws near, the pith-balls suspended from the conductor open wide, with either positive or negative electricity, and when the edge of the cloud is perpendicular to the exploring wire, a slow succession of discharges takes place between the brass ball of the conductor and one of equal size, carefully connected with the nearest spot of moist ground. I usually connect a large jar with the conductor, which increases the force, and in some degree regulates the number, of the explosions; and the two balls between which the discharges pass can be easily regulated as to their distance from each other by a screw. After a certain number of explosions, say of negative electricity, which at first may be nine or ten in a minute, a cessation occurs of some seconds or minutes, as the case may be, when about an equal number of explosions of positive electricity takes place, of similar force to the former, *indicating the passage of two oppositely and equally electrified zones of the cloud:* then follows a second zone of negative electricity, occasioning several more discharges in a minute than from either of the first pair of zones; which rate of increase appears to vary according to the size and power of the cloud. Then occurs another cessation, followed by an equally powerful series of discharges of positive electricity, indicating the passage of a second pair of zones; these in like manner are followed by others, fearfully increasing the rapidity of the discharges, when a *regular stream* commences, interrupted only by the changes into the opposite electricities. The intensity of each new pair of zones is greater than that of the former, as may be proved by removing the two balls to a greater distance from each other. When the centre of the cloud is vertical to the wire the greatest effect consequently takes place, during which the windows rattle in their frames, and the bursts of thunder without and noise within, every now and then accompanied with a crash of accumulated fluid in the wire, striving to get free between the balls, produce the most awful effect, which is not a little increased by the pauses occasioned by the interchange of zones. Great caution must of course be observed during this interval, or the consequences would be fatal. My battery consists of fifty jars, containing seventy-three feet of surface on one side only. This battery, when fully charged, will perfectly fuse into red-hot balls thirty feet of iron wire, in one length, such wire being $\frac{1}{270}$ of an inch in diameter. When this battery is connected with 3000 feet of exploring wire, during a thunder-storm, it is charged fully and instantaneously, and of course as quickly discharged. As I am fearful of destroying my jars, I connect the two opposite coatings of the battery with brass balls, one inch in diameter, and placed at such distance from each other as to cause a discharge when the battery receives three-fourths of its charge. When the middle of a thunder-cloud is overhead, a crashing stream of discharge takes place between the balls, the effect of which must be witnessed to be conceived. As the cloud passes onward, the opposite portions of the zones which first affected the wire come into play, and the effect is weakened with each successive pair till all dies away, and not enough electricity remains in the atmosphere to affect a gold-leaf electrometer. I have remarked that the air is remarkably free of electricity, at least more so than usual, both before and after the passage of one of these clouds. Sometimes, a little previous to a storm, the gold leaves connected with the conductor will for many hours open and shut rapidly, as if they were panting, evidently showing a great electrical disturbance.

"It is known to electricians, that if an insulated plate, composed of a perfect or an imperfect conductor, be electrified, the electricity communicated will radiate from the centre to the circumference, *increasing* in force as the squares of the distances from the centre; whereas in a thunder-cloud the reverse takes place, as its power *diminishes* from the centre to the circumference. First a nucleus appears to be formed, say of positive electricity, embracing a large portion of the centre of the cloud, round which is a negative zone of equal power with the former; then follow the other zones in pairs diminishing in power to the edge of the cloud. *Directly below this cloud,* according to the laws of inductive electricity, must exist on the surface of the earth a nucleus, of opposite or negative electricity, with its corresponding zone of positive, and with other zones of electrified surface, corresponding in number to those of the cloud above, although each is oppositely electrified. A discharge of the positive nucleus above into that of the negative below, is commonly that which occurs when a flash of lightning is seen, or from the positive below to the negative above, as the case may be; and this discharge may take place, according to the laws of electricity, through any or all of the surrounding zones, *without influencing their respective electricities* otherwise than by weakening their force, by removal of a portion of the electric fluid from the central nucleus above to that below; every successive flash from the cloud to the earth, or from the earth to the cloud, weakening the charge of the plate of air, of which the cloud and the earth form the two opposite coatings. Much might be said upon this head, of which the above is but a slight sketch."

Thus far had Mr. Crosse's observations gone on the phenomena of the thunder-cloud; but he was still uncertain how to account for the separation of the cloud into concentric zones. He thought long and earnestly on the subject, but no suggestion came; at length, one morning, while he was shaving, the explanation suddenly darted into his mind, and, with a schoolboy's glee, he shouted "Eureka!" and rushing down to his electrical room, with the lather on his chin, he immediately sought to test his theory by experiment. He was right; he *had* "found it." I give the explanation in his own words.

"I was first inclined to refer this phenomenon of the separation of the cloud into concentric zones to the fact of the cloud being a secondary conductor; and in consequence I insulated and electrified various plates of a *secondary* conducting power, such as moist wood, leather, &c., and compared the electricity which resided on every part of their surfaces with that which was communicated to insulated *primary* conductors, and I found no difference in the residence of the electric fluid, as in both primary and secondary conductors it radiated equally from the centre to the circumference; nor was the least symptom of an oppositely electrified zone discoverable. At last I hit upon what I believe to be the *real cause* of the phenomena. A cloud is, of course, a mass of vapour of a *secondary* conducting nature; but that is not sufficient to account for the zones. A cloud is composed of minute particles of water, each separated from its neighbour, and held in suspension by the caloric, which causes it to be elevated into the atmosphere in the form of vapour; consequently the whole cloud is *subdivided* into little conducting spheres, and resembles in this respect a dry plate of glass gently breathed upon, or a plate of glass dotted all over with spots of tin-foil. If you form a plate of this nature, and electrify the central spots with a spark from a charged jar, what is the consequence? Why, the communicated electricity will strike from the central spot across the contiguous spaces, and divide its electricity equally amongst them and in a circle; and when it has exhausted its *communicating* power, an *inductive influence* begins, which in its turn communicates the opposite electricity to the neighbouring spots, in a concentric circle, around the first nucleus formed. Here we have one pair of zones, which will, in like manner, be followed by a second pair, and so on, till the whole cloud is arranged accordingly; the central zones being the most powerfully electrified, and those at the circumference the weakest. By reasoning analogically, it *must* be so. The more powerfully electrified is the cloud, the wider and more extensive is each pair of zones, and also the more numerous. Should I be asked *what* influence is it that first impresses this electric power on the centre of the thunder-cloud? I could not presume to answer. Rudely speaking, evaporation seems to be the main cause. I should, in speaking of the conducting nature of clouds and vapours, make use of a new term, and call them *disseminated* conductors, in opposition to those of an uniform substance. It is the *interval*, the non-*conducting* interval between each particle of suspended water, which is the cause of these effects; it being a law of electricity that a number of small intervals between conducting substances impede the communicating power as much as one greater interval, and hence the inductive power."

In a private letter to Dr. Noad, on the same subject, Mr. Crosse observes: "With regard to your question as to exhibiting an *illustration* of the zones of a thunder-cloud, so as to be calculated for a public lecture, I have *often* and *often* thought upon it, but never succeeded to my own *perfect* satisfaction. I can, however, give you an account of an experiment which I hit upon many years since, the result of which was perfectly satisfactory, and which is closely connected with the electrical alternations in a thunder-cloud. The difficulty lies in the *exhibition* of the different states of electricity which exists in any *one* substance, in which the circumstances are such that no electrometer can be applied. I was asked by a friend how it was that a *chain* conductor was *inadmissible* on a ship, it being rent in pieces by such a streak of lightning as would fail to injure one *continuous* conductor formed of a much *smaller* diameter than that of which the chain was manufactured. My answer was, that the destruction of the chain was occasioned by the division between *each link*, which, however apparently small, occasioned during the passage of the electric stroke through the chain an inductive opposite electricity between every link, which increased with the length of the chain in such manner as to give an increased impediment to the even and uniform passage of the electric stroke, eack link becoming for *the instant* a receptacle of a greater charge than could possibly take place in one uniform rod. My friend's answer was, 'Can you *prove* this?' I replied, 'Were you an electrician, you would not ask me.' 'But,' said he, 'can you make this evident to my senses?' To which I answered that I would endeavour so to do when we next met. I afterwards reflected how this could be done, all electrometers being useless in such a case. I therefore thought of trying the old experiment on the projection of powders, and accordingly mixed an equal quantity by measure of red lead and sulphur, each in fine powder; a portion of this I threw into an old-fashioned powder puff, made to project the powder as from a pair of bellows. Now it is well known that if two metallic discs, furnished with insulating handles, be placed on a resinous plate, a few inches apart, and oppositely electrified by the knob of a Leyden jar, and then removed by their insulating handles, if the resinous plate be supported on its edge, and then the mixed powders be gently projected on the oppositely electrified spots, that the sulphur will separate from the read lead, and form a beautiful star on the positive spot, and the red lead will form on the negative in a star of a totally different appearance, somewhat resembling a circular ring of inverted pyramids, with their apices converging towards the centre. Thus are the two oppositely electrified spots rendered distinctly visible, though otherwise they are perfectly invisible. In this case of the chain, I resolved to endeavour to test its opposite electricity in a similar manner, and I took a chain composed of flat links, as thus —

about eight or nine inches long, and attached a silk thread to one end, for the purpose of insulation, and then laid the chain at full length along the middle of the resinous plate, after which I passed through it, for a few seconds, a positive electric current from an electrical machine, and then carefully lifted off the chain by the silk cord with a sudden jerk, so as not to create an alteration in its resident electricities, after which I set the resinous plate on its edge, and projected the mixed powder upon that portion of it on which the chain had laid, and had the satisfaction of finding an arborescent formation of sulphur in the shape of, and exactly under, the spot which had been occupied by each link of the chain, proving its *positive* electricity; and also in every interstice between each link I found two inverted pyramids of red lead, meeting at their extremities, proving the existence of *negative* electricity between each link; thus —

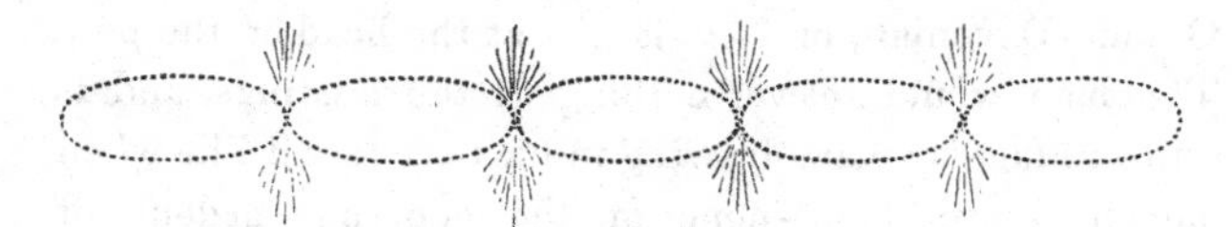

Great care must be taken, first, not to mix the two powders together till *just before using;* secondly, that the resinous plate be quite dry, and free from all electrical excitement, the best mode of doing which is to pass the surface of the plate over a spirit-lamp two or three times."

It is a curious contrast to turn to another phase of Mr. Crosse's character, to see the calm, patient philosopher, occupied in minute and long-continued investigations of Nature's laws, leave his laboratory, and ascend the hustings, where he would fling himself into the hottest arena of political excitement. With a temperament at once generous, vehement, and impetuous, he spoke outright whatever came uppermost in his mind, regardless of whether it might be politic or expedient. The following letter is eminently characteristic.

"183–.

"My dear Kenyon,

"I write in haste. We 're beat, almost dead beat. Money, cursed bribery, and intimidation of tenants, have beat us, added to our too great confidence and d—d mismanagement. Poor ——, our faithful and firm representative, is turned out, and a bigoted, narrow-souled Oxonio-Devonian, or Devilian, is at the head of the poll. This man would answer nothing at the hustings, and is quite a Machiavelian, drawing in his horns snail-like when danger approaches,—none of the generous ardour of youth and incautious nobleness of mind. Aristotle, which he must have studied, says, 'Youth does everything too much, age too little.' Our new member (may he soon be dismembered!) gives the lie to Aristotle, and while yet in his youth, excels in the dissimulation of age. He would have made an excellent cardinal in olden times, although he anathematises the Catholics. In fact, some are born before and some after the spirit of the age, and he, though young and inexperienced, is three centuries old. This I as good as told him on the hustings, where I seconded ——. I said nothing ungentlemanly, but I LASHED the two candidates and their supporters into no small rage. They luckily stuck the model of a church on a pole, and suspended a blue flag on it. This I pointed out to them, and told them we well understood the symbol, the house of prayer brought into a political contest. This was their religion, their morality, the name of all that was venerable and excellent, used as a cloak for worldly and tyrannical purposes; their bishops with incomes of secular princes, their working clergy with no incomes at all; in fact, that their real god was riches, power, rank, &c., and that it would be more honest to worship such openly, than under the mask of religious hypocrisy. Words to this effect I flung in their teeth, and hope to do so more effectually some other opportunity. * * * We are preparing for the next election, and are going to establish political unions, &c. &c. I have been driving to all parts of the country with —— ; our reception everywhere almost was *enthusiastic beyond description*,— the people are with us. Let me know a day or two before you come to Somerset. I hope to pass a couple of days with you at Lyme. My experiments at Broomfield go on famously. * * * *

"Yours ever,

"ANDREW CROSSE."

The period had now arrived when Mr. Crosse appeared before the public as a scientific man: few, perhaps, but those who knew him, could understand the distaste he felt to the *éclat* of science; he was not, however, careless of fame because he was indifferent to the good opinion of his fellow-creatures, but rather because he thought humbly of all human knowledge.

In the year 1836, the British Association for the Advancement of Science was to hold the annual meeting at Bristol. Mr. Crosse had never connected himself with any scientific bodies; he had passed a life of intellectual isolation with none around who were interested in his pursuits, none who would sympathise in his aspirations, or who could understand the nature of his experiments: of their full significance he was scarcely himself aware. Totally without ambition, he had for years worked on at intervals in the solitude of his own home; amidst the annoyances and vexations of life, he turned to philosophy as to a friend; he thought only of science as a means to exalt his own nature, and not as a means to raise his position in the world. Amongst the few who had long recognised Mr. Crosse's talents was Mr. Thomas Poole, of Stowey, the friend of Sir Humphrey Davy, and the Lake poets. This gentleman strongly urged him to attend the meeting at Bristol. "I was very uncertain about going," said Mr. Crosse, in speaking of the occurrence, "for I always shrank from pushing myself forward, and I was but little in spirits for such an occasion, for a constant succession of family illnesses had crushed me almost to the earth."

However, he went to Bristol, still without any idea of bringing forward his own discoveries. But he chanced to be dining at the house of a friend, and met Mr. Gilmar, Dr. Dalton, and some other scientific men, who were so exceedingly interested in the details which he gave them of his electrical experiments, that they begged him to bring forward an account before the chemical section, which he did, and afterwards repeated his statements before the geological section.

The enthusiasm with which Mr. Crosse's discoveries were received is almost beyond credence. The impression is vividly remembered even now, after the lapse of twenty years. When the circumstance is recurred to, persons, otherwise strangers, have described to me, in glowing language, the effect of words so simple, yet so earnest, which characterised Mr. Crosse's utterance of the results he had met with in searching into Nature's laws; the order of his inquiries was at once profound and original. The following extract is from a letter of Mr. E. L. Richards, published in the journals of the day. The writer, after detailing the observations of Mr. William Hopkins on dislocations of rocks, and their magnetic structure, and Mr. Fox's experiments on the influence of electro-magnetism on the formation of mineral veins, proceeds to describe "the magnificent discoveries of Mr. Crosse." He goes on to say, "That Mr. Fox's experiments proved the agency of a powerful principle in the formation and modification of metalliferous deposits hitherto little thought of, and evidently in strong connection with the causes so ably explained by Mr. Hopkins, is certain; but it remained for the discoveries so indefatigably followed up by Mr. Crosse, to exemplify how frequently and how materially this principle had been employed in the construction of some of Nature's most splendid productions. * * * It is difficult to give any just notion of the appearance of Mr. Crosse when he first got up to speak; simplicity and a perfect unconsciousness that he had anything extraordinary to communicate, were the prevailing features. In person he is tall, of light complexion, with a manner at once frank and open-hearted. His address was not polished; * * * he appeared that which indeed he stated himself to be, the child of seclusion, devoted to scientific pursuit, which had engrossed his mind for nearly thirty years, and forming one of the noblest instances on record of a man of wealth and station dedicating the best period of his life to the development of Nature's mysterious power, with the sole aim of benefiting mankind and doing honour to his country. Mr. Crosse's communication was unaccompanied by any apparatus, or other way of describing his discoveries, saving the use of the common section-board and a piece of chalk; yet his mode of explanation, his enthusiastic delivery, his earnest and solemn declaration of the means employed by Nature in the production of some of her most beautiful gems, wrought conviction in the mind of every one present; and when the results of his wonderful experiments were borne out by the personal observation of several eminent men, who readily attested the truth of his assertions, accompanied by the intimation that there was no production of the mineral world that would not probably be imitated by the ingenuity of man; the excitement became so great, and the applause so general, as to leave an impression on the minds of the dense mass that filled the lecture room scarcely to be equalled by any circumstance in their existence. At the commencement of his address he disclaimed any intimate knowledge of men or books; he declared he knew but little of mineralogy; that the whole course of his life had been devoted to one great object, the investigation of the theory of electricity, and that he had amused himself from time to time by an endeavour to imitate some of Nature's productions."

From the newspapers of the day is gathered the following somewhat bald account: —

Dr. Buckland, who occupied the chair at the Geological Section of the British Association, introduced Mr. Crosse in the following manner; after alluding to the interesting observations which had fallen from

Mr. Fox of Falmouth, he said: "There was also a gentleman now at his right hand whose name he had never heard till yesterday; a man unconnected with any society, but possessing the true spirit of a philosopher: this gentleman had actually made no less than twenty-four minerals, and even crystalline quartz (loud cries of 'Hear'); he, Dr. Buckland, did not know *how* he had made them, but he pronounced them to be discoveries of the highest order."

Mr. Crosse then described to the section, that by an arrangement in which he passed a voltaic current, excited by water alone, through certain mineral solutions, he had formed various crystalline bodies analogous to those found in nature. In these experiments, in which he used long-continued voltaic action of low intensity, he had obtained artificial crystals of quartz, arragonite, carbonates of lime, lead, and copper, besides more than twenty other artificial minerals. "One regularly shaped crystal of quartz, measuring $\frac{5}{16}$ of an inch in length, and $\frac{1}{16}$ of an inch in diameter, and readily scratching glass, was formed from fluo-silicic acid exposed to the electric action of a water battery from the 8th of March to the latter end of June, 1836." Mr. Crosse added, in conclusion, that he was fully convinced that it was possible even to make diamonds, and that at no distant period every kind of mineral would be formed by the ingenuity of man. "If," said he, "any members of the Association would favour me with a visit at my house, they shall be received with hospitality, though in a wild and savage region on the Quantock Hills, and I shall be proud to repeat my experiment in their presence." Mr. Crosse sat down amidst long continued cheering.

Professor Sedgwick then rose, and said he had discovered in Mr. Crosse a friend who some years ago conducted him over the Quantock Hills on his way to Taunton. The residence of that gentleman was not, as he had described it, in a wild and savage region, but seated amidst the sublime and beautiful in nature. At that time he was engaged in carrying on the most gigantic experiments, attaching electric wires to the trees of the forest, and conducting through them streams of lightning, and even turning them through his house with the dexterity of an able charioteer. Sincerely did he congratulate the section on what they had heard and witnessed that morning. The operations of electrical phenomena, instances of which had been detailed to them, proved that the whole world, even darkness itself, was steeped in everlasting light, the first-born of heaven. However Mr. Crosse may have hitherto concealed himself, from this time forth he must stand before the world as public property."

At the general evening meeting, the presidents of the different sections reported their proceedings, and Dr. Buckland alluded more particularly to the discoveries made by Mr. Crosse, which he said were such as were never before known. The patrons of science owed him an obligation for an achievement which would immortalise his name in the annals of geology. The president, the Marquis of Northampton, said that Dr. Buckland had observed that Mr. Crosse had no intention, when we came to Bristol, of communicating his discoveries, but was led to do so by the discussion he heard; this was a very singular and pregnant instance of the advantage derived from this Association."

At the Chemical Section, Mr. Crosse also read papers on "Some Improvements on the Voltaic Battery," and "Observations on Atmospheric Electricity." He had not brought with him from Broomfield any notes whatever of his experiments, so that he had to trust to his memory alone for the details, at once complicated and original. Mr. Crosse's discoveries had caused so much excitement, and were received with so much enthusiasm, that all manner of compliments were quickly showered down upon him. Dr. Dalton did him the honour to say, that he had never before listened to anything so interesting. This remark of the great chemist, I know, pleased him much, for he felt it was the honest and sincere appreciation of a true lover of philosophy. But while his name was being bruited about with honourable mention, his discoveries canvassed and applauded (not entirely however without detraction), and invitations and compliments passed upon him, he was only anxious to be away: to use his own characteristic words, "I slipped away out of it all," and two or three days before the British Association had closed their meeting at Bristol, Mr. Crosse was back to the wild Quantocks, encased once more in the solitude of his own home, devoting himself to the isolated pursuit of his beloved science, and shrinking, absolutely shrinking, from the celebrity he had acquired.

The following extract is from a letter of John Kenyon, Esq., to Andrew Crosse. They had been schoolfellows together, and remained attached friends through life.

"Dear Crosse,

"* * * I had absolutely made an engagement to come to the Bristol meeting, but thinking that I should meet no very particular friends, I begged off again, though I was at perfect liberty to move: now I could bite off my finger ends, to think that I have missed you and Poole. Having been on the wing for a dozen days, I had read no detailed account of the goings-on of the British Association meeting; but yesterday I had a glorious (a vain-glorious) letter from Poole (but he was vain of *you*), written expressly to relate your triumph; he rightly calculated what pleasure it would give me. '*Cum populis frequens,* lætum theatris tu crepuit sonum.' He has told me what extraordinary applause you received, and how modestly you retreated from being one of the great show-beasts of the meeting; and I, who have known you forty years, neither wondered at the success nor the modesty. He tells me, too, how perspicuously and yet eloquently, and with what little (or no) divergence you spoke,—'dicendi recte sapere est et principium et fons,' and you have the 'provisam rem;' and I am more particularly pleased because he tells me that your 'observers' at Broomfield who came to see were satisfied. Well, we must hear more of you: please God that the family affliction which (perhaps) you are still suffering, in the health of your son Richard, may pass away, and leave you at liberty to give a quickened attention to these things. But I do you an injustice when I say quickened attention—I mean a sustained attention. No man had ever less need to be quickened by vulgar feelings of fame or profit than yourself; no man has ever been more completely satisfied to go on alone. 'By those pure eyes, and perfect witness of all-judging love,' I and other of your friends will be proud for you, if you will not consent to be proud for yourself. * * Well, one of my plans is, that you must come and spend a fortnight with me in London. * * We will see a thousand things, and see a thousand things together. * * I know how desirous you are to see me at Broomfield, but you have a sick house. For many years all your cares, administrative and others, have been there,—as you once said to me, 'I have a stake through my body which nails me to Quantock.' * * How could we have believed, forty years ago at Seyer's, conjuror as we called you, that you should be writing with crystal pens of your own making (referring to the electrical crystals of quartz which scratched glass), and I be carrying your autographs about the country. Well, it has pleased God to try you with much household distress for some years—in your own unsatisfactory health, in the health of Mrs. Crosse, in the health of living children, and of those who are gone, and with other trials to which, more or less, flesh is heir to; but you have also been blest with compensation in a mind (as you once said with your characteristic energy) that, with a magnifying glass, could be happy in a prison, looking at a straw, and in a resisting buoyancy of spirit; and now, after all your furnace-watchings and explosions, whether vinous or electrical, comes at last a deserved reputation in part payment. * * * If you were not aware what Becquerel had done, so neither, it appears, were any of the Geological Section: you came forward, assuming no comparative merit over any man whatever, and, in fact, did not come forward, but were brought forward by accident. * *

"I am truly and affectionately yours,

"JOHN KENYON."

The following extracts are from an account of a visit to Broomfield,* which appeared in a local journal shortly after Mr. Crosse's appearance at Bristol.

"If, when you come to the village of Kingston, about three miles from Taunton, you turn upon your right into a dark and narrow lane, you will soon find yourself climbing with toil a difficult and very steep hill; the road is rough, and the hedges meeting over head give it an aspect of the profoundest gloom. But by day, in the summer time it is deliciously cool and shady, and a very wilderness of wild flowers. Having conquered this hill, a turn of the road on your left conducts you to a park adorned with many fine beeches, on one side of which you behold a sheet of water, with a shrubbery in the background, whose very aspect invites you to trespass in it. All this you see as you walk under a row of trees that overshadow the road; and if you are a stranger to the place and its owner, you will wonder what can be the meaning of the mast-like poles fixed at the tops of the loftiest trees, by which a line (so it appears) is carried round the park till it is lost in the shrubbery. Presently you see a mansion, oddly roosted in a hollow, under the

* By Edward W. Cox, Esq.

ridge of the high ground you are treading, just as if the soil on which it had been built had suddenly sunk on some fine morning; for it is difficult to believe that an architect could have placed it there on purpose. It is a plain building on the outside, but it contains that within which passeth show. Knock fearlessly at the door; the votaries of science are always welcome there. Your name? your station? your calling? your property? Trouble not yourself about any of these things, nor hope thus to commend yourself to the inmates. You are a *man*, you have a *mind*, you venerate *science*, even if you know little of it: these are your passports into the mansion. Are you a stranger? you will not be so long. 'One touch of nature makes the whole world kin.' In ten minutes you feel as if you had known your kind and generous host twenty years. Your entertainer attracts you, and you gaze, as much as you politely may. He is now in his velvet jacket, his laboratory costume; his frame is made for activity; light but muscular, having not an ounce of superfluous fat, with a trifling stoop at the shoulders; his face too is thin and long, with a fine forehead, a well-shaped nose, and pointed chin. Its expression is highly intellectual, with an air of seeming melancholy, which is in fact one of thought; but a lengthened gaze discovers in it a lurking propensity to fun, which continually peeps out at the corners of his eyes and in the curl of his lips. His hair is brown partially silvered by age, which is betrayed only there; for his gait and countenance have all the liveliness and energy of youth; his step is springy, his voice cheerful, his aspect that of one who enjoys good health, and its attendant good spirits: such, dimly outlined, we must confess is the personal appearance of ANDREW CROSSE.

"Had you never before heard that name, or if you had not known that you were about to visit one who had distinguished himself in the pursuits of science, you would soon discover that you are in the company of a man of genius, that you are conversing with one who has thought for himself, and refused to subject his mind to the chains of authority, and to bow before the *dicta* of schools. The presence of genius you discover in Andrew Crosse before you have conversed with him for a quarter of an hour. The talk of most men, even of those who are reputed as wise or witty, is merely a repetition of that which you have heard, in substance if not in form, from other men fifty times before, and read as often. But Mr. Crosse's talk is his own. You may differ from his opinions, you may question his accuracy, you may contest his arguments, you may smile sometimes at views that may seem to you visionary and wild, because they are different from your own habitual trains of thinking, and therefore startle you; but you cannot complain that they are commonplace; they are not echoes of the voices of others, nor gems in a new-setting — *alter et idem* — stolen from books old or new. * * Particularly striking is Mr. Crosse's eloquence, when he tells you the wonders of his favourite science of electricity, of its mysterious agencies in the natural phenomena of the heavens above, of the earth beneath, and of the waters under the earth; how it rules alike the motions of the planets and the arrangement of atoms; how it broods in the air, rides on the mist, travels with the light, wanders through space, attracts in the aurora, terrifies in the thunderstorm, rules the growth of plants, and shapes all substances, from the fragile crystals of ice to the diamond, which it makes by toil continued for ages in the womb of the solid globe. As he describes to you all these wonders, not imaginations of a dreamer, but realities which he has himself seen and proved, by producing, by the same agent and the same process, only in a lesser degree, the same results, his face is lighted up, his eyes are fixed upon the ceiling, present things seem to have disappeared from him, lost in the greater vividness of ideas which his full mind throngs before him ; he pours out his words in an unfailing stream : but, though he has a command of epithets, he finds language inadequate to express his conceptions of the might of that mysterious element which, though so very mighty that it could annihilate a world as easily as it lifts a feather, he has summoned from its throne, compelled into his presence, guided with his hand, and made to do his bidding ! — thus surpassing the fabled feats of the enchanters of old.

"Before you visit the hall where this mighty power is at work day and night, obedient to his command, and daily showing itself in some new shape (a very Proteus), yet unable to escape from the potent spell of the magician by whom it is compelled, you would like to stroll with your host into the plantations and gardens. Step through the window on to the lawn and follow him. But, beware! you are no longer in the company of a sage philosopher, but of a man (we might almost say of a boy) full of fun and frolic, and laugh and joke? That roguish twinkle of the eye and half suppressed curl of the lip betoken mischief. Look at him! There is not a trace of the student in his manner or in his talk. Can this be he whom we heard but two minutes since discoursing, with the rapture almost of inspiration, of the mysteries of science? He is as merry now as a child at play. What a glorious laugh, a real, honest, hearty laugh!—not a stifled titter, as if he were ashamed to be natural. What a step and jump, as though age had been worsted in wrestling with him, and had succeeded only in frosting his hair with its breath in the struggle! * * * * * * Mr. Crosse talks as one who has discovered how large a portion of the current coin of learning, which even scholars boast of as real wealth, is, in fact, base metal that will not endure the test of reason and experiment; and so, instead of being proud of what he has done, he is humbled by it. He has learned by years of study the last, most difficult of all lessons—*not to know*. He rates himself much lower than any other person rates him—a rare phenomenon! Nor is this humility of mind assumed; he has not a spark of affectation; the conviction is in him and it shows itself.

"But to proceed now into the penetralia of the mansion, the philosophical room, which is about sixty feet in length and upwards of twenty in height, with an arched roof,—it was built originally as a music hall,—and what wonderful things you will see. * * A great many rows of gallipots and jars, with some bits of metal, and wires passing from them into saucers containing some dirty-looking liquid, in which, with much attention, you may espy a few crystals. * * * It was the invention of a battery by which the stream of the electric fluid could be maintained without flagging, not for hours only, but for days, weeks, years, that was the foundation of some of Mr. Crosse's most remarkable discoveries. * * * Crystals of all kinds, many of them never made before by human skill, are in progress. * * * But you are startled in the midst of your observations, by the smart crackling sound that attends the passage of the electrical spark; you hear also the rumbling of distant thunder. The rain is already plashing in great drops against the glass, and the sound of the passing sparks continues to startle your ear. Your host is in high glee, for a battery of electricity is about to come within his reach a thousandfold more powerful than all those in the room strung together. You follow his hasty steps to the organ gallery, and curiously approach the spot whence the noise proceeds that has attracted your notice. You see at the window a huge brass conductor, with a discharging rod near it passing into the floor, and from the one knob to the other, sparks are leaping with increasing rapidity and noise, rap, rap, rap — bang, bang, bang; you are afraid to approach near this terrible engine, and well you may; for every spark that passes would kill twenty men at one blow, if they were linked together hand in hand, and the spark sent through the circle. Almost trembling, you note that from this conductor wires pass off without the window, and the electric fluid is conducted harmlessly away. On the instrument itself is inscribed in large letters the warning words,

"'Noli me tangere.'

Nevertheless, your host does not fear. He approaches as boldly as if the flowing stream of fire were a harmless spark. Armed with his insulted rod, he plays with the mighty power; he directs it where he will; he sends it into his batteries: having charged them thus, he shows you how wire is melted, dissipated in a moment, by its passage; how metals—silver, gold, and tin—are inflamed, and burn like paper, only with most brilliant hues. He shows you a mimic aurora, and a falling star, and so proves to you the cause of those beautiful phenomena; and then he tells you, that the wires you had noticed, as passing from tree to tree round the grounds, were connected with the conductor before you; that they collected the electricity of the atmosphere as it floated by, and brought it into the room in the shape of the sparks that you had witnessed with such awe. And then, perhaps, he will tell you that the electricity lies in a thunder-cloud in zones, alternately positive and negative, and he will add that he is able at all times thus to measure the electrical state of the atmosphere; and he will tell you many curious facts which he has subsequently observed relative to that state at various periods of the day and night, and at the different seasons of the year. * * * * * *
And as you talk with Mr. Crosse, in rambling fashion, flying from theme to theme, and gleaning something new and instructive from all, you will probably (for how can thinking minds avoid it?) touch upon the lofty topic of religion,

"'Of reason and foreknowledge, will and fate;'

you will then discover that Mr. Crosse is, like almost all philosophers, a profoundly religious man, and yet he has not escaped the old slander which has assailed all men of science, imputing to them atheism and infidelity. An atheist, indeed! Look at him! hear him! His every look and tone and word show his sense of a present Deity; veneration of his Creator's greatness, love of his goodness, reliance upon his providence, are habitual with him. He is a Christian in the best and noblest import of the name. But he abhors priestcraft when directed to purposes of human aggrandisement and power. * * * * * But while you have been thus pleasantly and instructively engaged in discourse, the sun has set, the summer twilight even has vanished. Unwillingly you depart; you bid your generous host good-night, and wish him health and happiness with your whole soul, again and again repeating it, as you walk homewards, enjoying the cool night air, and thinking over what you have heard and seen; you discover more of interest in the sky, and in the stars, and in the trees dimly outlined, and in the solid earth you tread, than ever you had recognised before; your mind has been enlarged and refined, your heart has been softened, your good principles have been strengthened, your prejudices have been shaken, if not overthrown; your notions of the Creator and His works have been enlarged, your passions have been calmed, the shadows which social follies and vices had cast upon your natural feelings and temper have been melted away, and you are altogether a wiser, better, a nobler being — more of a man — for this your day spent with ANDREW CROSSE."

I remember an anecdote connected with the arrangements of the atmospheric conductor in the organ gallery which caused some amusement at the time. The servants were always desired to avoid touching any of the apparatus, but it appears that a housemaid, who was carrying on her vocation of dusting, went up and touched the brass cylinder bearing the words "Noli me tangere." There was a considerable amount of electricity present in the atmosphere, and she got a rather severe shock. She forthwith went to her master, and complained that "That nasty thing in the gallery had nearly knocked her down." "I thought that I told you never to touch the apparatus," said Mr. Crosse. "Yes, sir; but I thought you had written 'No danger' on it!" If all bad translators were so corrected, it would save the world a great deal of literary trash.

There was another story which we often used to laugh over. A large party had come from a distance to see Mr. Crosse's experiments and apparatus. He had been taking them to different parts of the house, as was his wont, explaining his various philosophical arrangements: at length, on arriving at the organ gallery, he exhibited two enormous Leyden jars, which he could charge at pleasure by the conducting wires, when the state of the atmosphere was sufficiently electrical. An old gentleman of the party contemplated the arrangement with a look of grave disapprobation: at length, with much solemnity, he observed: "Mr. Crosse, don't you think it is rather impious to bottle the lightning?" "Let me answer your question by asking another," replied Mr. Crosse, laughing: "Don't you think, sir, it might be considered rather impious to bottle the rain water?"

A Chinese proverb wisely and quaintly says, "Towers are measured by their shadows, and great men by their slanderers." It is quite curious to observe the amount of ignorant and absurd abuse that was showered upon Mr. Crosse from some quarters. To jealousy and prejudice may be ascribed much of the calumny of mankind: if there is one class of the community who enthusiastically accept a new truth, there are always a number of oppositionists who seek to pull down the idol that their fellows have raised. By some it was stated that M. Becquerel's experiments on electro-crystallisation were altogether antecedent. It is certain that Mr. Crosse commenced in his laboratory these imitations of nature's workings before the year 1820, and it is equally true that the results were not brought before the public till 1836. The two philosophers had been working simultaneously somewhat on the same class of experiments, each without any knowledge of the other, and I feel quite assured that neither felt any but the most kindly and generous sentiments, in respect to their mutual and kindred discoveries, when the results of their labours were brought into juxtaposition. I have in my possession a letter from M. Becquerel to Mr. Crosse, couched in terms of great appreciation and full acknowledgment of the important discoveries made by the English electrician: and every one who knew Mr. Crosse personally must be well aware how pure was his love of science, and that to him the development of truth was far beyond the exaltation of himself or any other individual, and he of all people was the least likely to feel jealousy, or to desire to raise his own reputation at the expense of another.

A number of very ridiculous attacks were made on Mr. Crosse, which for the most part he treated with the silent contempt that they deserved. When dispassionate and rational objections were raised to any particular results or course of experiments made by him, no one was more ready than Mr. Crosse to treat with respect and attention the suggestions of intelligent and unprejudiced opposition. The following letter, however, was written with something of the "*honest indignation*" which formed a prominent feature in Mr. Crosse's character.

"*To the Editor of the Atlas.*

"Broomfield, near Taunton.
"Jan 31st, 1837.

"It is exceedingly disagreeable to me to be compelled to bring the name of so unimportant an individual as myself before the public, and had I not met with such an unfair attack as that of Dr. Ritchie, I should not have presumed to make the following statement. From my boyhood I have detested nothing so much as cant or humbug of any sort; and if I had been disposed to form a high opinion of myself, the very severe lessons I have met with in my passage through life would have been amply sufficient to humble me. It was by mere chance that I attended the meeting of the British Association at Bristol, having only made up my mind so to do on the day previous to the commencement of the business. So far was I from having the least idea of making a communication, that I previously feared lest, from the pressure of the crowd, I should lose the opportunity of hearing what was going on. Chance led me into conversation with some eminently scientific men, who, having heard some observations of mine, requested me to make them public. To this I assented, and what followed is known. I never attached the least merit to what I had done, noı *tacked* the word *discovery* to any of my experiments, but gave a simple statement of what took place. It was for the public to form what opinion they pleased. At the close of my observations before the chemical section, *I stated distinctly*, that if any one present wished to question me, I would endeavour to answer them to the best of my ability; and Dr. Ritchie, who was present at the Association, and knew what was going on, might have taken an excellent opportunity of exposing my ignorance, face to face. No questions were asked. M. Becquerel's experiments were then brought forward, of which I had not seen any account, although I heard shortly before the meeting of his having formed sulphurets of lead and silver by the electric action. I stated that my experiments were made from the mere love of science, and that I did not wish to detract from or presume to set myself in competition with any one. The only statement I myself published since the meeting, was contained in your paper, in a letter in answer to one received from you. A great many mistakes and exaggerations concerning me have been circulated in different papers at different times; but it is no fault of mine that such things occurred. In fact, it cannot be expected that those who have not made a particular science their study should be correct in their details concerning such science. It is very true that, in my electrical room, a brass ball connected with an atmospherical conductor is suspended over a battery, so arranged as to be united or disunited at pleasure; but such battery is a *common electrical*, and most assuredly *not a voltaic one*. Dr. Ritchie must have known well that the absurd report of having brought into my house streams of lightning as large as the mast of a ship originated in a joke that fell from the eloquent lips of a distinguished professor who attended the meeting. Dr. Ritchie, however, in supposing such a miracle possible, compares the *inferiority* of such an enormous electrical current with the *superiority* of that manifested by Richman and Romas, the former of whom was unfortunately killed by want of proper precaution, having contrived an apparatus to bring the electric fluid into his house, but apparently without making a due arrangement to carry it out. The latter, by means of a kite, elevated far above the highest of my poles, brought down an amazing quantity of electricity; but such a temporary apparatus as a kite is ill-calculated for scientific purposes, to say nothing of the extreme danger attending it. I have really no wish to be knocked on the head in aspiring to eclipse my neighbours. Dr. R. next proceeds with something more tangible, — my letter to you. He is first offended with the term circular batteries. It is quite clear that plates answer as well as cylinders; but I made use of the latter shape on account of my employing common glass bottles with their necks off as insulators, into which I fitted two zinc and copper cylinders.

He next insinuates that I claim as a discovery the filling the cells with water instead of dilute acid. When did I state this? I may, however, in justice to myself observe, that I have not heard of effects produced by other batteries filled with common water at all equal to what are produced by the arrangements I adopt, and for the truth of this I appeal to those scientific men who have witnessed their effects. Dr. R. then notices my remark on the greater power of these batteries between the hours of *seven and ten* in the morning, and that such increase of strength was unconnected with any metre whatever. Here he observes that I do not know the properties of the agents used in a voltaic battery, and proceeds to lecture me on the alteration of the conducting power of primary and secondary conductors by the influence of heat, summing up with the inference that the increase of temperature is the cause of the increase of power observed, and that, too, between the hours of seven and ten in the morning! I shall only say, in reply to this, that my batteries are placed in a room with a south aspect, and that every fair day the sun shines full upon them, at which time the shock they give to the human body is *decidedly less* than at the hours alluded to. As to my ignorance of the effects of heat, &c., on voltaic batteries, I have tried fluids of all temperatures, and often used boiling water in the cells, which occasions a considerable increase of power. I do not agree with Dr. Ritchie that 800 pairs of plates used with water are equivalent in power to 50 pairs of the same size, and under the same circumstances, used with dilute acid. Of course the former are far superior in intensity, but they are by no means equal in the density of the current. Dr. R. next finds fault with my stating that an increase of number of plates produces more than a corresponding effect in power; and he then adverts to his investigations published in 1832. I stated at Bristol that my apparatus was incomplete, and that I was in the midst of a train of experiments. I am so still, and it is most unfair to expect a perfect result during the course of investigation. I have not seen the account of those experiments to which Dr. R. alludes, but if they were performed with batteries each cell of which was not separately and carefully insulated, it would amply account for his failing to produce effects equal to what would otherwise have taken place. To say the truth, I have made a much greater number of experiments in this branch of the science than Dr. R. may be aware of, but am not about to enter at present upon the comparative powers of the simple electric, the decomposing, the fusing, or the magnetic effects produced from differently formed batteries. *Opinion to me is but a wind, experiment a rock.* Now comes his attack on my electrical crystallisations of substances. I before stated all I knew concerning M. Becquerel. Had M. Becquerel been the Englishman and I the foreigner, I do believe that gentleman would have received from Dr. R. the censure instead of myself, *as if it were a crime for a countryman hitherto unknown to experimentalise at all.* The next remark on me is: 'According to Mr. Crosse, either pole of the battery will crystallise equally well.' When and where did I state this absurdity? It is the first time that I have heard of it. I have, however, met with some curious and quite unexpected facts which bear on this part of the subject, which neither chemist nor electrician would have expected without previous trial. Then follows: 'The crystallisation of quartz and carbon is still doubtful.' That I have produced the first by the electric action long continued, as also arragonite, I can prove by unimpeachable witnesses; the last (carbon) I have not attempted. As I never aimed at procuring public applause (although I am deeply sensible of the great kindness I have received from my friends and the public) but have pursued science for its own sake, those shafts of Dr. Ritchie fall powerless against me. I cannot, however, refrain from adding that I would scorn to admit for one instant such a spirit towards another as he has evinced towards me, a stranger to him in all respects, save public report, even for the power of crystallising carbon.

"I am, in the meantime, sir,

"Yours sincerely,

"ANDREW CROSSE.

"P.S. I should have sent this answer long since, but have been prevented by severe illness. I must beg in future to decline engaging in scientific warfare with any one, having neither inclination nor time for that kind of amusement."

A few months after the meeting of the British Association at Bristol, where Mr. Crosse's name had been so prominently brought forward, another circumstance took place, which caused intense excitement in connection with one of those mysteries which have hitherto, and probably ever will defy the explanations of philosophy. The world was startled by the publication of an account of very unexpected appearances in some of Mr. Crosse's electrical experiments. Insects, in fact, were found to have been developed under conditions usually fatal to animal life, namely, in highly caustic solutions and out of contact of atmospheric air. The circumstance certainly was extraordinary, and deserved, as it still does deserve, further inquiry. The true spirit of philosophy lies in an equal mean between credulity and denial; but the prejudices acquired by education rarely allow us to observe the just medium. Mr. Crosse was no entomologist or physiologist, therefore he did not pretend to know whether these little animals, which had so strangely presented themselves between the poles of the voltaic circuit, were a new species or not (and it is still an open question, though they clearly belong to the *Acarus* tribe); he knew that under certain arrangements he could reproduce them at pleasure, and that unless these conditions were observed they did not appear. He formed no theory in respect to their development, and he was far too honest to attempt an explanation of what he freely allowed he did not comprehend. Without any idea of a formal announcement, Mr. Crosse stated the fact to some private friends: he chanced, however, to name the matter in the presence of the editor of a West of England paper, who immediately, *unauthorised*, but in a very friendly spirit, published an account of the experiment; which account quickly flew over England, and indeed Europe, satisfying at once the credulity of those who love the marvellous, and raising up a host of bitter and equally unreasoning assailants, whose personal attacks on Mr. Crosse, and their misrepresentations of his views, were at once ridiculous and annoying. One gentleman actually wrote to him, calling him a "disturber of the peace of families," and "a reviler of our holy religion." Mr. Crosse's answer was very characteristic: after disavowing all intention to raise any questions connected with either natural or revealed religion, he went on to observe that he was sorry to see that the faith of his neighbours could be overset by the claw of a mite. He was also accused in one of the public prints by some anonymous individual of being the cause of a blight which took place at that time. The poor little insects, the so-called "*Acarus galvanicus*," had had "greatness thrust upon them." It was certainly not the intention of the experimenter that they should "make themselves great." I believe Southey was the first person to whom Andrew Crosse related the circumstance of their appearance. The latter was walking over the Quantock Hills, as was his wont, with his eyes fixed on the ground (a habit acquired from mineralising), and pondering with amazement on the strange development of what he had expected to be crystals into living animals. Thus reflecting on the result of his experiment, he met Southey toiling up the hill behind a carriage which was to convey him to Mr. Poole's at Stowey. The poet and the philosopher were acquainted, and most friendly was their greeting. Andrew Crosse, full of the subject occupying his thoughts, at once communicated the fact of the curious appearance he had met with,—holding Southey fast to hear the most minute details of the experiment, as the "Ancient Mariner" might have held the wedding-guest. "Well," said Southey, "I am the first traveller who has ever been stopped by so extraordinary an announcement." It had been frequently remarked that there was a very strong personal resemblance between Mr. Crosse and Southey, and certainly it appeared so from the portraits and busts that I have myself seen of the latter: the great difference was that the electrician's brow was the most massive, and the lower part of his face less correct and not so chiselled, but indicating in the breadth of jaw more power than that expressed in the striking countenance of the poet. In coming to the Quantock Hills, Southey was returning to the scene of old associations: he, together with Wordsworth and Coleridge, had, in the early part of the present century, lived near Stowey, and many are the poetical reminiscences, as we all know, that they have left of this favoured spot. I think I have heard Mr. Crosse say that he was acquainted slightly with Wordsworth; Coleridge, I believe, he never saw, but his brother, Richard Crosse, himself a great metaphysician, called on Coleridge on one occasion, and left him after three hours' talking in the middle of his second sentence! Among the celebrities who had visited the Quantock, was Sir Humphry Davy, also the guest of Mr. Poole of Stowey. It was in the autumn of the year 1827 that a circumstance occurred, which has been slightly noticed in Dr. Davy's life of his brother, and which I well recollect hearing Mr. Crosse describe. Sir Humphry Davy was staying with Mr. Poole: he was then weakened by the dire disease that was so soon to end fatally, and with difficulty he had driven as far as Fyne Court with his friend. Mr. Crosse took them over the house, and at length into the laboratory. "Never shall I forget," said Mr. Crosse, "seeing Davy's fine melancholy eyes brighten up, as he looked at the furnaces. For a few moments he seemed himself again, the languor of disease had fled, and his old activity was expressed in every look and action;" but he was passing away, it was the beginning of the end. There is something profoundly melancholy in reflecting that but a few years have passed, and all, all are gone,—the philosophic Davy, the philanthropic Poole, the warm-hearted, the enthusiastic Andrew Crosse. It is well for us that hope gives, what life cannot, a future.

But to return to the subject from which I have unwittingly digressed, namely, the "electrical insects." The following is a letter from Andrew Crosse to Harriett Martineau, in answer to some inquiries made by that lady, prior to her publication of an account of the acari, in the "History of the Thirty Years' Peace." The order of subject has been preserved rather than that of date.

"15. Charles Street, Manchester Square, London.
"August 12th, 1849.

"Madam,

"Your communication of August the 3rd is now before me. I shall endeavour to reply to your questions in such a manner as they deserve. Allow me in the first place to state that I have not the slightest objection to your dealing as you please with this answer of mine. You are welcome to publish it if you think proper, or thrust it into the fire, where many of those kind commentators on some of my experiments would gladly have thrust me. It is the bounden duty of philosophical men not to reject or admit as fact, any assertion, without close and fair investigation. This would save a world of trouble, and be of the highest importance to the science concerned.

" Ever since I have enjoyed the faculty of thinking, two feelings, apparently somewhat opposed to each other, have been predominant in my mind,—the first exciting within me an ardent wish of knowing more, and the last causing a conviction that the utmost extent of human knowledge is but comparative ignorance. Feeling as I have done the whole of my life, it is not likely that I should plume myself upon any imaginary successful results of a course of experiments, or that I should presume to lay down a theory upon so mystical and perhaps unapproachable a subject as to the origin of animal life.

" As to the appearance of the acari under long-continued electrical action, I have never in thought, word, or deed, given any one a right to suppose that I considered them as a creation, or even as a formation, from inorganic matter. To create is to form a something out of a nothing. To annihilate, is to reduce that something to a nothing. Both of these, of course, can only be the attributes of the Almighty. In fact, I can assure you most sacredly that I have never dreamed of any theory sufficient to account for their appearance. I confess that I was not a little surprised, and am so still, and quite as much as I was when the acari made their first appearance. Again, I have never claimed any merit as attached to these experiments. It was a matter of chance. I was looking for silicious formations, and animal matter appeared instead. The first publication of my original experiment took place entirely without my knowledge. Since that time, and surrounded by death and disease, I have fought my way in the different branches of the science which I so dearly love, and have endeavoured to be somewhat better acquainted with a few of its mysteries. Now, suppose that a future son of science were to discover that certain novel arrangements should produce an effect quite contrary to all preconceived opinion, would this discovery, however vast it might be, humanly speaking, be such as to stir up in a mind properly constituted an inferior sense of the omniscience of the Creator? It is really laughable to anticipate such a result, which could only be engendered in the brains of the enemies of all knowledge.

"In a great number of my experiments, made by passing a long current of electricity through various fluids (and some of them were considered to be destructive to animal life), acari have made their appearance; but never excepting on an electrified surface kept constantly moistened, or beneath the surface of an electrified fluid. In some instances these little animals have been produced two inches below the surface of a poisonous liquid. In one instance they made their appearance upon the lower part of a small piece of quartz, plunged two inches deep into a glass vessel of fluo-silicic acid, or, in other words, into fluoric acid holding silica in solution.

A current of electricity was passed through this fluid for a twelvemonth or more; and at the end of some months three of these acari were visible on the piece of quartz, which was kept negatively electrified. I have closely examined the progress of these insects. Their first appearance consists in a very minute whitish hemisphere, formed upon the surface of the electrified body, sometimes at the positive end, and sometimes at the negative, and occasionally between the two, or in the middle of the electrified current; and sometimes upon all. In a few days this speck enlarges and elongates vertically, and shoots out filaments of a whitish wavy appearance, and easily seen through a lens of very low power. Then commences the first appearance of animal life. If a fine point be made to approach these filaments, they immediately shrink up and collapse like zoophytes upon moss, but expand again some time after the removal of the point. Some days afterwards these filaments become legs and bristles, and a perfect acarus is the result, which finally detaches itself from its birth-place, and if under a fluid, climbs up the electrified wire, and escapes from the vessel, and afterwards feeds either on the moisture or the outside of the vessel, or on paper or card, or other substance in its vicinity. If one of them be afterwards thrown into the fluid in which he was produced, he is immediately drowned."

(Here follows an account of an experiment which is detailed in a note at the end of the volume.)

"Conflicting opinions have existed, as to whether the

acarus developed under the above circumstances be of a new description or not. I know not whether this may be of any consequence, as it is very easy for so minute an animal to have escaped particular observation ; and besides, I have observed a variety amongst the acari so produced; but this I leave to entomologists. I have never before heard of acari having been produced under a fluid, or of their ova throwing out filaments; nor have I ever observed any ova previous to or during electrisation, except that the speck which throws out filaments be an ovum; but when a number of these insects, in a perfect state, congregate, ova are the result. I may now remark that in several of these experiments fungi have made their appearance, and in some cases have been followed by the birth of acari. In one instance a crop of fungi was produced upon the upper end of a stick of oak charcoal, plunged into a solution of silicate of potash, kept negatively electrified for a considerable time, and covered by a bell-glass inverted over it in a dish of mercury. The charcoal before being used was taken red-hot from a fire. There is evidently a close connection between animal and vegetable life: but one thing is necessary to be observed, that such experiments as those I have just touched on must be varied in every possible form, and repeated over and over again with unflinching perseverance, and with the most sharp-sighted caution, in order to attain satisfactory results.

In conclusion, I must remark, that in the course of these and other experiments, there is considerable similitude between the first stages of the birth of acari and of certain mineral crystallisations electrically produced. In many of them, more especially in the formation of sulphate of lime, or sulphate of strontia, its commencement is denoted by a whitish speck: so it is in the birth of the acarus. This mineral speck enlarges and elongates vertically: so it does with the acarus. Then the mineral throws out whitish filaments: so does the acarus speck. So far it is difficult to detect the difference between the incipient mineral and the animal; but as these filaments become more definite in each, in the mineral they become rigid, shining, transparent six-sided prisms; in the animal they are soft and having filaments, and finally endowed with motion and life. I might add much more to the above sketch, but it would be more fit for a pamphlet than for a letter.

"With great respect,

"I am, Madam, yours truly,

"ANDREW CROSSE."

This letter has anticipated considerably, in point of time, that portion of the narrative of which this chapter is devoted to the consideration.

The succeeding three years after the meeting of the British Association at Bristol, in 1836, was a period of great domestic trial to Mr. Crosse. Family illness of a peculiarly afflicting character kept him almost a prisoner to his own roof, and distracted his mind with griefs and anxieties that none could feel more keenly than himself. On the sad occasion of a bereavement a friend thus writes to him: — "God bless you, my dear Crosse! — your life has been one of severe visitations, but God has given you the blessed power of attaching your friends to you, and you cannot suffer without many a heart's feeling for you." Mrs. Crosse's health was much weakened by many years of suffering, and his dearly loved brother was becoming gradually more and more of an invalid. It was Mr. Crosse's habit, for a long period, to spend every Sunday with him. He lived at a cottage of his own erection, about three miles from Fyne Court. He had married early in life, but had no family. The strongest possible attachment existed between the brothers. Their conversations were very generally upon metaphysical subjects. Mr. Crosse was not informed upon this science himself, but he adopted the opinions of his brother, whose profound investigations and exalted sentiments had the most favourable influence on the mind of one whose studies were entirely of a physical and experimental character.

The subjoined extracts from letters, written at this period, tell something of the way in which time went on at Broomfield, but the truest and saddest part of a man's life is often that which it is not the business of a biographer to narrate.

"To ——

"Broomfield, May 4th, 1832.

"My dear Sir,

"I received your letter the day before yesterday. The kindness of it I shall never forget. I am in the midst of all sorts of business, — selling hay, barking oak, cutting down poles, gardening, &c. &c. Far above all, I am working like a slave in my laboratory, and have two fires *constantly* burning night and day. I have formed crystals on several new plans, and I am preparing a very extensive apparatus. I wish my means were half as ample and extensive as the apparatus I would fain construct! I do not go to London to see the gewgaws and frippery and childish nonsense of the coronation. How much mankind have to learn before they begin to be ashamed of such trash!

I have just put together a water battery of sixty-three large zinc and copper cylinders, each cylinder equal to a nine-inch square plate. It gives a small but intense constant stream of light, between two charcoal points the heat of which will fire gunpowder. I am about to increase it to 100 pairs.

"Five thousand of such cylinders as these would make a glorious exhibition, but they would cost 500*l.* Each pair of cylinders is contained in a glass jar, which holds about three pints. The shock through the thin part of the skin, even quite dry, is almost intolerable. It is my belief that 1000 of such cylinders, or even less, would produced potassiun from alkali. I am half stewed with the heat of my furnaces, which I am obliged to watch closely. * * * * *

"Yours truly,
"ANDREW CROSSE."

"To W. H. Weeks, Esq.,
"Sandwich.

"Broomfield, near Taunton,
"August 17th, 1840.

"My dear Sir,

"Most happy am I to find from your very friendly letter, that I have in any way contributed to your happiness, even for a short period. I have gone through so many trials myself, that I am glad to be enabled to contribute to the comfort of another, and more particularly a friend of Lettsom. * * *

"On Saturday night last, at twelve o'clock, in walking home from Taunton, I saw a splendid aurora borealis, which occupied a tolerably large portion of the atmosphere. The moon was shining brightly in the south-east, and the aurora in the north-west, in the form of a series of pillars of crimson fire, which shone through the dark rolling masses of cloud in a very extraordinary way. It appeared like a series of alternate pillars of flame and dark vapour. Over head was a very distinct mackerel sky, with a stratum of rolling dark clouds far below. Between these two strata the light of the aurora struck out all at once in the form of pyramids of pale fire, shooting to the zenith.

"The pale light of a moon three parts full contrasted with the electric pyramids and crimson pillars of fire, mixed up with the rolling masses of cloud, gave to the scene a *wild* and *unearthly* appearance. I was greatly impressed by it. Since then we have had torrents of rain and furious winds; the clouds are electric but not highly charged.

* * * * * *

"ANDREW CROSSE."

Relative to the subject of atmospheric electricity, I subjoin a letter of Dr. Buckland the eminent geologist, addressed to Mr. Crosse.

"Oxford, Nov. 5th.

"My dear Sir,

"I have just now (and it is but just in time for your purpose) fallen upon the chapter in Arago's *Annuaire* for 1836, in which he mentions the phenomena of falling stars observed in late years between the 10th and 15th of November.

"I consider his theory at p. 293., which refers them to little asteroids, to be quite chimerical, and that your explanation, which would refer these phenomena to electric agency in our atmosphere, will prove to be the true one.

"* * * * * It is I think most important to get *crystals* of *quartz* of as large a size as possible. I should like to know the date of the quartz crystal with which you made scratches on my watch glass. I, in common with every other person who feels an interest in the progress of all magnetic discoveries, am looking forward with intense interest to the result of the manifold experiment you are now conducting, the bearing of which extends over some of the most important problems in physics and physiology; and I shall be most happy to be the medium of laying before the Geological Society of London any memoir you may have the goodness to prepare connected with the influence of electricity on individual minerals, or on the condition of the earth.

" And remain, my dear sir,

" Always very truly yours,

" W. BUCKLAND."

" To W. H. Weeks, Esq.,
" Sandwich.

" Broomfield, near Taunton,
" August 1840.

" My dear Sir,

" I received your two letters with much pleasure. You ask me in the first a question as to the *quality* of the electricity in different parts of the country in a similar state of weather. In answer to this I may state that in *fair weather* without clouds the electricity is *invariably positive* all over the world; I think I may safely say this positive state is greater or less, according as the evaporation is greater or less. Its *maximum* is at *sun-rise* and *sunset*, its *minimum* at *mid-day* and *midnight*. I observe that the first precursor of a thunder-storm is generally a diminution of the usual quantity of electricity in the atmosphere. This is succeeded by the *panting* or constant opening and shutting of the leaves of a gold leaf electrometer when connected with the atmospheric conductor. Then follows the storm, &c., greater or less, as may be. After the storm nature appears exhausted, and little electricity is visible. Rain is generally negative. Thunder clouds, snow, hail, fogs exist in alternate zones of positive and negative in general. Sometimes a snow cloud *possesses no zone*, but only when small. Sometimes no electricity can be obtained from a fog, on account of the insulators being wet. The aurora borealis, falling stars, and common summer sheet lightning do not affect the exploding wire.* * * * * *

"Yours truly,
"ANDREW CROSSE."

"To W H. Weeks, Esq.,
"Sandwich.

"Broomfield, near Taunton,
"Sept. 13th 1842.

"My dear Sir,

"I am always much pleased at the sight of your handwriting, morally and scientifically, and I should never allow a post to intervene, without answering your letters, did a kinder fate allow me sufficient time; but alas! this is not the case, as we seem to have little in our power, and are tost about like withered leaves before the gale. It is only in my power to wish, and I wish you all happiness in this world, and humbly trust we may meet in a better, which I look forward to with not an atom less hope, from the denunciations of those who style themselves orthodox, to such unbelievers as myself.

"It is very true that I have serious thoughts of going to the Continent for three years from the time of my starting, which cannot be before the spring of 1843. From the time of my birth to the present hour I have ever been far too careless in money transactions, and my love for science has not only led me to a considerable expense *directly*, but has, by calling nearly the whole of my attention elsewhere, prevented me from looking into my affairs with that scrutiny which is absolutely essential in a country establishment, and in the management of landed property. I have therefore, of course, been cheated tremendously, which I have to thank myself for. My great expense has been in building, and in repairs on farm houses, in which I have suffered immense imposition. Although I am MUCH ATTACHED to this place, yet in common prudence, I must quit it for a time, and carry on my experiments on a more limited plan, and in a cheaper country than this. I am happy to say that my family will all be well provided for, and for myself it is of little consequence as to the increase or diminution of a few comforts, as I am very easily satisfied, and my life cannot extend to a much longer term. The income tax has put the finishing stroke to my determination; but still I look forward to returning to my home at the expiration of three years. * * * I have hit upon a new plan, and a much more efficient one of forming my electrical crystals. I have succeeded in forming the carbonates of *lime*, *strontia*, and *baryta*, in *water alone*, well crystallised in a very short time. Carbonate of *strontia*, $\frac{1}{10}$ of an inch long, within a week. I am now trying that most insoluble substance *sulphate* of *baryta*, and can just see some minute crystals forming in *water alone!* * * *

"Yours truly,

"ANDREW CROSSE."

Mr. Crosse's intention of leaving Broomfield for the purpose of retrenchment was for a time laid aside, and finally abandoned. Domestic illness continued to have increasing claims upon his time, which, together with many other sources of annoyance and grief, weighed him down almost to the very earth. In January of the year 1846, Mr. Crosse had the great misfortune to lose his beloved wife and brother, who died within four days of each other.

CHAP. IV.

LABOURS IN THE LABORATORY.

On crystallisation.—Separation of metals from their ores by electricity.—Purification of sea water, &c.—Electricity applied to medical purposes.—The perforation of non-conducting substances by the mechanical action of the electric fluid.—Electro-vegetation.—New mode of making impressions in marble.—Metals dissolved in distilled water.—Blood kept fluid.—Analysis of the voltaic battery.—The mine in the garden pot.—Mechanical action.

FOR the sake of perspicuity, the details of Mr. Crosse's experiments on crystalline formations are condensed into the following arrangements, which consist principally of his own memoranda, either noted down at the time, or written afterwards from memory. The experiments themselves were conducted at intervals, through a series of years; but it must be observed that the most important results were arrived at long before they were made public.

As it has been already noticed, his first experiment on the water of Holwell Cavern was made about the year 1807, in which he obtained crystalline carbonate of lime at the negative pole of the voltaic battery. A large proportion of the other factitious crystals named were the results of experiments made about the year 1817 and the succeeding years. Mr. Crosse's memoranda of that period make mention of his "forming mineral substances by slow electric action." As experience became his teacher, he improved and simplified many of his voltaic arrangements, and he was enabled to ensure, in many cases, absolute results.

Believing the laws of form to be subject to certain molecular arrangement caused by electric attraction and repulsion, Mr. Crosse experimented on the combination of substances under voltaic action, *anticipating* thereby the formation of crystals. The results of his experiments showed that he was not mistaken. But at the same time, though he attributed the laws of form to electricity, he never denied that electricity itself possibly is the *secondary* cause only. Some *one* original law may regulate the different forces. Mr. Crosse explained this matter in a public lecture in the following familiar manner: —

"Light, heat, magnetism, and electricity are qualities of matter which are intimately connected with each other. One great law may regulate them all, and produce them all but as yet we are ignorant of it. I might compare them to four bottles filled with different fluids boiling over a fire, the same cause — the fire — producing the boiling of all."

"*Ars longa, vita brevis,*" were constantly the words which Mr. Crosse would repeat with a sigh, as he looked with disappointment at the small progress made in his imitations of nature. But the real motto of his laboratory was, "It is better to follow nature blindfold than art with both eyes open." This expression explains the character of his mind, and the manner in which he sought results. When he walked out he read, not in the book of man, but in the book of God. His acute powers of observation would reveal to him some peculiarity in the organisation of plants or combination of mineral substances, which often proved the first suggestion for a train of interesting experiments. Mr. Crosse ever evinced the most wonderful patience in his scientific arrangements; for months, even for years, he would wait for results, and watch the slow induration of what he hoped might be an agate, or the minute aggregation of crystals, whose slowly developed facits he would carefully note down from time to time. At an early period of his experiments on crystalline formations, he was not unfrequently disappointed from the fact of his having employed too strong an electric action. He used to say, "You cannot hurry nature;" too rapid an action throws down the substance in an amorphous state; atoms seem only to assume a crystalline form when they have time to arrange themselves in a state of polarisation to the surrounding atoms.

It had been the received chemical law that "to effect crystallisation there must be perfect *rest;* the arrangement of the particles must not be disturbed by motion." But in speaking of crystallisation, Mr. Crosse says: —

"There are other conditions besides electricity necessary to be observed, such as more or less even temperature, absence of light, and in many cases CONSTANT MOTION of the fluid holding the crystallisable matter in solution, either by dropping from the roof of a cavern, or by water constantly flowing, or by the continual elevation and depression of the surface of the subterranean waters; which surface is for ever varying, low in summer, or more or less overflowing in winter; but *constantly in motion.* It is this eternal motion that greatly facilitates the growth of crystals. This would seem a strange doctrine in the chemical laboratory, where *perfect rest* is more or less essential to the formation of well-defined saline crystallisations; but such is by no means the case with metallic and earthy matters." He says: "I have kept up a constant electrical action, for three successive months, upon fluids in a state of unceasing ebullition, in a sand heat furnace, day and night, without a moment's rest, the evaporated fluid being duly watched in the most careful manner. Yet the crystals formed were as perfectly solid and regular as similar ones taken from a mine, and were much accelerated in their growth, both by the heat employed and by the motion communicated by that heat. * * * Now there are two reasons why heat and motion are greatly conducive to electrical crystallisation. The first occasions a more rapid evaporation of the water holding the crystallisable matter in solution, and causes the fluid to be a far better conductor of electricity.

"*Motion* so disposes the atoms that it agitates, that, being polarised by the electric action, that is, each atom having its opposite extremities rendered positive and negative, they present themselves more readily to the opposite pole of the battery,—the negative end of the atom to the positive pole, having its outside still positive, or the positive end of the atom, as the case may be, being drawn to the negative pole, having its outside still negative; so that there is no impediment to the even and quiet passage of the electric current and the continual transfer of atoms acted upon to their respective poles. This may be better understood by placing a common magnet flat upon a table, with a sheet of paper lying upon it. If you let fall a mass of iron filings at once below the paper, they will be attracted to the magnet upon it, into the form of a rude misshapen heap; but if such filings be slowly sifted through a fine sieve, they will assume the form into which they are attracted by their respective polarities, and present a beautifully regular appearance, in obedience to the forces of the magnetic current. There is another condition essential to the production of nearly, if not quite all, regularly metallic, and most earthy crystallisations. It is the interposition of a *porous medium* between the two opposite electrical poles engaged in the work of forming minerals. In art this is brought to pass by the intervention of tabular surfaces, or cups of porous earth, or other porous material, which is used to separate the fluids or substances acted on, so as to bring them together slowly and regularly into a solid form. * * * We have thus seen that a union of *electric action* with a moderately *uniform temperature*, and *sufficiency of heat* to prevent congelation of the fluid under action, absence of light, together with the interposition of a more or less *porous medium*, will attract the crystallisable matter from its solution, and produce a variety of forms, which will not make their appearance without such conditions. We have likewise seen that those crystallisations or formations are greatly assisted by constant motion. * * Under these circumstances, I have produced about 200 varieties of minerals, exactly resembling in all respects similar ones found in nature, as well as some others never before discovered in nature or formed by art. A specimen which I have made of a *subsulphate of copper* is an instance of this. Still, there are a vast number of minerals, which, in the present state of the science, defy the ingenuity of man to imitate, but many of which have been produced by *central* or *volcanic heat,* or *immense pressure* added to the other requisites. The last thing to consider is, from what source does the required electric action arise? Now, in answer to this, as far as we know, it most probably arises from one of the following causes: first, from terrestrial electric currents, caused by permanent magnetic action passing at right angles to them; or secondly, from similar electric currents excited by the union of vast strata of dissimilar rocks in contact with subterranean waters; or thirdly, from similar currents, either excited or aided by a central or volcanic heat, perhaps coming under the laws of thermo-electricity; or fourthly, and lastly, by *local electric action.* The following details exemplify what I mean by the latter, viz., local electric action. Some years since, being at Weymouth, I observed some rounded limestones and some sea-shells embedded in the clay of a small perpendicular cliff, each stone and shell being covered with crystals of sulphate of lime. On looking around to investigate the cause of the formation of sulphate of lime upon these substances, I discovered a stratum of decomposing sulphuret of iron, running horizontally on the top of the cliff and just below the soil; accordingly, I reasoned thus:— The rain water penetrating the soil moistened the sulphuret of iron, and decomposed it; the oxygen of the water converting the *sulphuret* to the *sulphate;* and the sulphate of iron, being a soluble salt, passed through the clay, and was slowly admitted into contact with the surfaces of the limestones and sea-shells. A local electric action was excited, in which the limestones and sea-shells became negative, whilst the upper stratum of sulphuret of iron was positive. The sulphate of iron and carbonate of lime suffered each a decomposition, and sulphate of lime was produced in a crystallised form upon the negative surfaces of the limestones and shells, carbonic acid gas being liberated: moreover, this iron being deprived of its sulphuric acid, absorbed oxygen, and was converted into the red oxide of iron, which was abundantly precipitated around the base of the crystals of sulphate of lime, in a powdery form.

"In order to prove the correctness of this theory, on my return home I took a large basin, half filled it with pipe-clay, which I kneaded up with water to the consistence of moist putty, and embedded in the clay some pieces of limestone and some sea-shells. I next formed a stratum upon the clay of powdered sulphuret of iron, and then filled the basin with common water, and put it aside in a dark cupboard for a twelvemonth. At the end of that period I brought it into the light and examined it with no small anxiety; but was delighted to find that every piece of limestone and sea-shell which had been embedded in the clay, when taken out, washed, and dried, was covered with prismatic crystals of sulphate of lime exactly similar to those found in the cliff at Weymouth; but of course they were small, though perfect. Such are the effects of LOCAL ELECTRICITY. Observe that here no battery was used, nor metal in a metallic state. It was simply a close imitation of nature, but followed out only for a year; whereas nature has at her command unlimited time and resources.

"In the formation of crystalline matters from their solutions, and more especially from such as are somewhat concentrated, I have often met with a great impediment in a *counter electrical action*, which increases as the power of the battery diminishes.

"This effect is analogous to the action of Ritchie's secondary pile, and is to be accounted for on exactly the same principle. I have no doubt but that this counter action must necessarily exist in the natural world, and may be the principal cause which limits the growth of crystals to their usual dimensions. I have long believed with others that what is termed chemical affinity is nothing but a greater or less electrical attraction. . . . I think it highly probable that earthy crystals in great variety, as well as metallic, may be formed without applying artificial electricity in a direct way, or making use of any metal whatsoever, but simply by interposing a porous diaphragm between two solutions containing crystallisable matters having an affinity for each other, or, in other words, an electrical attraction, which, being kept moist, allows the slow union of the substances which produce crystallisation. If this view be correct, it is evident that a solid substance which is decomposable may be substituted for one of these solutions; or two solid substances capable of chemically acting on each other by the addition of water. In either of these cases crystals would be the result."

Mr. Crosse has thus answered the remark that crystalline formation is effected "*without* electricity," as the mode of expression is, viz., without *applied electricity*. And as regards the crystals formed by evaporation, may not the ascending vapour be in one state of electricity, say *positive*, and the substance or solution be *negative*, by induction? Electricity is so universally present, that it is impossible to effect any change, whether in the organic or inorganic kingdom, without, in a greater or less degree, exciting this mysterious law.

The following details of experiments on crystallisation are given as nearly as possible in the order in which they occur in Mr. Crosse's own memoranda:

"Holwell Cavern is remarkable for having its roof and sides covered with crystals of arragonite. The water of this cave contains ten grains of carbonate of lime and a minute portion of sulphate of lime in each pint. I took some of this water and filled a wine-glass with it, and by means of two platina wires connected it with the poles of a voltaic battery, composed of 200 pairs of five-inch plates, in Wedgwood troughs, filled with *water alone*. During the first nine days no alteration took place at either pole, except the usual disengagement of the gases at either end. On the tenth day I perceived, by the aid of a lens, a deposit of crystalline carbonate of lime at the negative pole: at the end of about three weeks the whole of the carbonate of lime contained in the glass was arranged in a crystalline form around the negative wire. These crystals were mostly irregular, but some of them were rhomboids. I then removed the wires and found by a test that no lime remained in the water. The positive wire was unaltered. Having reason to imagine that the experiment would succeed better in the dark, I arranged in my cellar forty pairs of two-inch single plates, plunged in *separately insulated* gallipots, and passed a constant current of electricity by means of platina wires through a pint and a quarter of the Holwell water in a glass basin. At the end of six days the crystalline deposits took place at the negative pole, and at the expiration of about two months the whole of the carbonate of lime contained in the water was deposited. I should mention that the negative wire was coiled round a piece of common limestone, taken from the same cavern, and that the crystals were first formed upon the wire, and next upon the limestone : some of these crystals were rhomboidal. Having next considered that the carbonate of lime in the cavern was formed with access of air, I contrived an insulated filter, through which I passed factitious water of the same nature as that of Holwell Cave, but much more strongly impregnated with lime. This water I caused to fall in successive drops for weeks together on a piece of porous insulated brick, each end of which was connected by means of platina wires with the poles of 100 pairs of five-inch plates, in cells filled with common water. The brick rested on a glass funnel, and was exposed to the action of the electric fluid, which was readily conducted through it on account of its being constantly wet. A crystalline but much finer deposit took place at the negative pole, occupying a much larger space then before; but as I grew impatient, and expected more rapid results, I removed the glass funnel with the brick and supported it over a glass cylinder filled with the same solution, into which the lower half of the brick was plunged. The arrangement was continued two or three weeks longer, when to my great surprise crystals, apparently of arragonite, in a needle form, arranged themselves in the form of a circle around, but not touching, the positive pole: these crystals spread over the nearest portion of the glass funnel, and partially cemented it to the brick. I next resolved to repeat these experiments on other substances, and formed a battery of eleven large insulated cylinders, from the poles of which I passed a dense stream of the electric fluid by means of platina wires through a similar piece of brick half plunged in fluo-silicic acid. Here the most remarkable phenomena took place. As this acid commonly contains a small portion of lead, in the course of a few hours this lead was deposited at the negative pole in an arborescent form, and at the termination of three weeks very brilliant crystals of what appeared to be silica tipped the ends of a portion of the *branches* of *lead.* These crystals were of indetermined forms, very minute, and perfectly transparent. In the course of a few days they wholly disappeared, and silex began to form at the negative pole in the form of mammillated chalcedony, while the positive wire was covered with a chocolate-coloured substance first noticed by Sir H. Davy. While this action was going on, the fluo-silicic acid was continually evaporating, carrying with it a considerable portion of silica into the atmosphere. So as to save expense in these experiments, I made one of the copper wires tipped with platina wire, to which they were firmly bound by a fine brass wire. It was curious to observe the silicious crystals which were formed by the evaporation of the fluid on the brass negative wire. These were translucent, and arranged in irregularly crystallised concretion. Having a strong suspicion that the impurity occasioned by the presence of the lead gave a wrong direction to the silica, I repeated this experiment, and at the termination of three weeks carefully removed all the lead formed at the negative pole just previous to the formation of silica at that pole. In consequence, in the course of a few days, the silex was *determined* to the POSITIVE pole, in an arborescent form, but not at all resembling chalcedony. I let the action continue some weeks longer, during which, although the evaporation continued as before, not one particle of silica was carried into the atmosphere, but remained concentrated in the solution, until one or two hexagonal prisms, terminated by hexagonal pyramids, stood out from the lower part of the brick. These crystals, when they began to form, grew rapidly; and at the end of a few days from their formation I removed one and examined it, but was much disappointed to find it too soft to scratch glass. Accordingly, I changed my plan, made a solution of silex in aqueous potash, and exposed a piece of porous brick half plunged in it to the action of 100 pairs of two-inch plates in water; then I had to wait three weeks before any change took place. At the end of that time a deposit of silica surrounded the positive pole in a white opaque mass, no change taking place at the negative. A few days after this, a most extraordinary appearance took place: in the middle of the brick fifteen or twenty hexagonal figures, in the form of outlines, made their appearance, like No. 1. These outlines appeared to be composed of silicious matter, and were visibly elevated above the surface of the brick. In a few days the angles of each of these hexagonal figures were connected with the centre by means of similar lines, like No. 2. And after a further interval of a few more days, lines parallel to the east side of the hexagon filled up the figure, like No. 3.

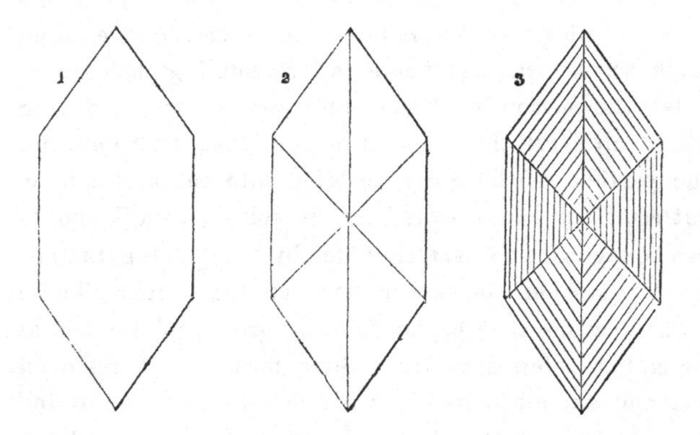

These figures, when viewed obliquely, were evidently silicious, and many of them partially raised an hexagonal pyramidal summit above the surface of the brick.

Then came another disappointment. I had the most sanguine hopes that these figures, some of which were one sixth of an inch in diameter, would gradually fill up, and give me the unspeakable pleasure of viewing from beginning to end Nature's slow, but exquisitely beautiful process of the formation of a crystal, step by step. But something was wanting ; — other crystals, mostly irregular, shot up on all sides, disfigured the beautiful beginning, and left me to contrive new arrangements. Still the specimen formed is exceedingly beautiful. * * * I have repeated similar experiments on a considerable variety of other substances : amongst others I have made a solution of nitrate of silver and copper, into which I half plunged a piece of porous brick, and connected it with 100 pairs of five-inch plates. The bottom of the vessel was strewed with pounded green bottle-glass (the action has been kept up many weeks): at first arborescent silver was formed at the negative pole ; then very round perfect beads of green carbonate of copper or malachite studded the middle of the brick ; next the pounded glass was decomposed, nitrate of potash crystallised round the edges of the vessel, and a portion of the silex of the glass combined with the oxide of copper, forming chrysocolla, somewhat mixed up with the silver at the negative pole. The positive wire for many weeks was surrounded by a line of demarcation forming a circle of nearly three quarters of an inch in diameter, within which, till lately, the surface of the brick has been unaltered. At present it presents the appearance of a very perfect semicircle of arborescent silver, which half surrounds, but does not touch, the positive wire.

"In making an experiment with a view to form the triple sulphate of lead, antimony, and copper, or bournonite, I obtained an unexpected result. I took certain proportions of sulphate of copper, sulphate of lead, and sulphate of iron, and white oxide of antimony, mixed them intimately together, and added three times their weight of pounded glass. I used this latter substance to steady the platina wires, and I poured water on the whole, subjecting it to the action of thirty pairs of two-inch plates. First metallic copper was precipitated on the negative wire; above it, a minute portion of metallic lead. In four days' time cubic crystals of brilliant iron pyrites arose from the decomposition of the sulphate of iron, which almost entirely covered the surface of the copper; a large portion of the lead was converted into yellow massicor, the positive pole during the whole process rapidly separating oxygen. I have also formed phosphate of copper at the positive pole, by plunging copper wires in solution of phosphate of soda, and I am forming acicular crystals of carbonate of lead, by keeping a plate of lead electrified positively, in opposition to one of copper both plunged in water. By making a coil of copper wire, and opposing it to a plate of zinc under water, and exposing the two to the action of 100 pairs of five-inch plates, the copper coil was covered with a very beautiful coating of mammillated carbonate of zinc.

"NOTES OF EXPERIMENTS.

"*An Account of some Experiments, in which Electric Currents of moderate Intensity are employed to effect a Transfer of a Portion of one Substance to that of another, through the Medium of common Water.*

"*Experiment* 1*st.*—I took a common red brick, and placed it horizontally, with each end resting on a half brick of the same sort, in a large glazed pan filled with the spring water of the country. This water is comparatively pure for spring water, but contains a small quantity of chlorine in combination with other matters. A wire of platinum of about one thirtieth of an inch in diameter was passed twice round one end of the brick, about three quarters of an inch from the extremity, and connected with the positive pole of a sulphate of copper battery; and a similar arrangement connected with the other end of the brick with the negative pole. The pan was placed very evenly on a smooth wooden table, and the apartment kept dark. Shortly after the battery was set in action on the brick a strong smell of chlorine proceeded from the vicinity of the positive wire. I should have mentioned that the upper surface of the brick stood half an inch above the level of the water, and was covered with clear river sand, carefully sifted on it to the thickness of one third of an inch.

"I occasionally added fresh spring water to that in the pan, to keep it at the same level. At the exact termination of a year I took apart the apparatus, and found as follows:—On attempting to lift the whole brick, from the two half bricks that supported it, I found that while the positive end was easily removed from the brick below it, the negative end required some little force to separate it from its support; and when the two were wrenched asunder, I found that they had been partially cemented together by a tolerably large surface of beautiful snow-white needle crystals of arragonite, thickly studding that part of the brick in groups, the crystals of each radiating from their respective centres. Although I had found this interesting mineral in small particles, yet I had never obtained the fiftieth part of the quantity I procured in this experiment, which covered a surface of several superficial inches, both on the lower surface of the upper brick at its negative end, and on the upper surface of the half brick that supported it. Here and there were formed, in some of the little recesses in the brick, elevated groups of needle arragonite, meeting together in a pyramidal form in the centre, while in the open spaces between were some exquisitely formed crystals of carbonate of lime, in cubes, rhomboids, and more particularly in short six-sided prisms, with flat terminations, translucent and opaque, sufficiently large to determine their form without the use of a lens. On emptying the water from the pan, I found at its negative end, at the bottom, a very large quantity of snow-white carbonate of lime, to the extent of some ounces in weight, in the form of a gritty powder in minute crystals. The lime in these experiments was extracted from the brick at the positive end, assisted materially by the minute quantity of chlorine naturally existing in the spring water, forming a chloride of calcium, which, as it was constantly decomposing, as constantly gave back the same atoms of chlorine, to attack a fresh proportion of calcium at the positive end. Thus a very minute quantity of chlorine, subjected to a long succession of decompositions, when in a state of chloride of calcium, was sufficient to produce the most extensive effects.

"*Experiment 2nd.* — I set in action a sulphate of copper battery of six pairs of plates, on a piece of crystallised carbonate of strontia, between two pieces of clay slate, in a glass jar of spring water, — the strontia connected with a positive pole, and the lower slate with the negative by platinum wires. The effect has been remarkably great; the whole of the negative wire under water, and nearly the whole of the lower slate, being covered with pearly-white carbonate of strontia in a botryoidal formation in considerable quantity. The upper slate is less covered with the same mineral; but upon this are some definite crystals. The interior of the glass is partially covered with stalactitic carbonate of strontia; this appearance after six months' action. The platinum wires coiled round mineral substances, forming the passage of the electric circuit, are generally observed to have eaten out a bed, often to the depth of upwards of half an inch, making a sort of neck round the stone, in fact.

"*Experiment 3rd.* — I set in action, with a battery the same power as the last, and exactly in a similar manner, a piece of carbonate of baryta positively electrified, the lower slate being negative. The negative wire under the water and a portion of the lower slate is covered with a beautiful mammillated formation of carbonate of baryta, formed in less quantity but of larger size than the strontia, and of a more pearly-white colour.

"*Experiment 4th.* — I set in action, with a sulphate of copper battery of six pairs of plates, a piece of solid opaque white quartz, suspended in a glass basin, by a platinum wire kept positive, and in the same a similar piece of quartz, kept in the same manner, — negative, — with other pieces of quartz between the two, the basin being filled with a solution of pure carbonate of potassa not concentrated. A long time elapsed before the slightest change was visible, but at length a number of minute crystals of quartz, perfectly transparent, are formed upon the positive stone. I am inclined to believe that the same effect would be produced, but in a considerably longer time, *without the use of an alkali.*

"*Experiment 5th.* — Crystals of sulphate of barytes and muriate of copper formed after some months' action, on porous pan in tumbler, by arc of zinc and copper. Muriate of barytes in pan, and copper in solution; sulphate copper in tumbler.

Experiment 6th. — Set in action arc of zinc and copper wire, round sulphuret of antimony, in porous pan and basin. Zinc in dilute sulphuric acid, and copper, &c., in concentrated solution of sulphate of iron ; for about four months. This formed magnetic oxide of iron on the sulphuret of antimony. N.B. It was placed in the magnetic meridian.

"*Experiment 7th.*— Set in action arc of zinc and copper

in glass cylinder and porous pan in the dark. Zinc in the pan, and nitrate of strontia; copper in sulphate of copper in cylinder. Crystals of sulphate of strontia were formed on the outside of the porous pan a quarter of an inch long. Time, one year and one week.

"*Experiment* 8*th.* — Phosphate and carbonate of copper formed at the positive pole on copper, by plunging two copper plates in solution of phosphate soda, and connecting them with either pole. Some months' formation."

(Probably a portion of *carbonate* of soda was found in the phosphate, which is rarely procured genuine.)

Experiment 9*th.* — Copper perfectly crystallised in octahedrons on a plate of copper, six pairs of plates (zinc and copper); in tumblers with common water; copper in porous pots with sulphate of copper. Not two months' action."

Mr. Crosse made various experiments on electro-crystallisation with the addition of heat. The solutions acted on were placed in a sand-bath and kept at boiling heat for several weeks together.

"*Experiment* 1*st.* — May 11th, three o'clock. Set in action arc of zinc and copper in porous pan and basin. Zinc on zinc (plate five inches square); copper on two copper lumps, weighing 1 lb. 7 oz. 60 gr. Troy. Zinc in basin and muriate ammonia. Copper in pan and sulphate of copper. Answers *capitally.* Took abroad, May 28th, beautifully covered with crystals, red oxide of copper, mostly, and metallic ditto : weight of lumps, 2 lbs.; gain, 5 oz. 60 gr. This solution was kept at boiling point the whole time. Perfect octahedral crystals of metallic copper were formed by these experiments.

"*Experiment 2nd.*— Set in action under similar conditions, large arc of zinc and copper, in large porous cylinder and basin; zinc on zinc in hydrosulphuret of potash; copper on three lumps of copper, weighing 2½ lbs troy and 100 grains, in sulphate copper, with arc of copper wire. When taken abroad, the lumps beautifully covered with crystallised copper and *subsulphate* of copper, and some red oxide: gained in weight, 7 oz. 70 gr. in 23 days 17 hours. In these experiments the various solutions were occasionally fed with their respective salts, and constantly supplied with water. The evaporation amounted to 7 gallons *per diem.*"

Mr. Crosse effected the crystallisation of quartz in several different modes. It will be remembered that the fact of producing quartz crystals artificially was acknowledged by the elder Becquerel as resulting from Mr. Crosse's labours. Quartz crystals have appeared in the following arrangements: —

"Set in action four pairs of plates of sulphate of copper

and water; or an enclosed solution of dilute silicate of potash, with porous pot containing the solution made *negative*, and placed in a glass basin of the same solution, in which was suspended a piece of white quartz made positive with platinum wire. On the edge of the glass basin, close to the positive wire, are formed numbers of prismatic crystals of quartz."

In an experiment in which Mr. Crosse made use of fluo-silicic acid he obtained a quartz crystal, a hexahedral prism measuring $\frac{3}{16}$ of an inch in length by $\frac{1}{16}$ in breadth; it readily scratched glass.

In 1837 Mr. Crosse thus writes: —

"I have lately been making some interesting experiments on the injurious effects of light in electrical crystallisation. Of this, which I have long suspected, I am now convinced. Among other experiments, I filled a large glass jar with lime-water, and suspended in it a coil of stout copper wire, and put it away in a dark place. In a few months the wire was covered with brilliant crystals (hexahedral) of carbonate of lime, $\frac{1}{16}$ of an inch in length. I then removed the jar, &c. into the light, and in the course of about six weeks the crystals *entirely* disappeared.

"I have made a new formation containing lead, copper, and zinc, combined with a very large proportion of sulphur, forming a supersulphuret, in most brilliant crimson prismatic crystals."

From the previous memoranda it will be seen that Mr. Crosse obtained the following crystallised minerals: —

Iron. — Crystallised iron pyrites ÷ *
Copper. — Crystallised in cubes.
" In octahedrons.
" Crystallised arseniate.
" Crystallised red oxide in octahedrons.
Lead. — Crystallised carbonate ÷
" Crystallised sulphuret.
Gold. — Crystallised.
Silver. — Crystallised oxide of.
" In octahedrons.
Silex. — Crystallised.
Lime. — Fluate of, in cubes.
Arragonite.
Sulphur. — Transparent crystals of ÷
Barytes. — In tabular crystals.
" In needle crystals.

In the course of these experiments he also obtained mineral substances in the form in which they occur in nature, though not in a crystallised condition. In the following lists, moreover, there may be some

* This mark means that the mineral has been formed by others, and there is an uncertainty as to whom the priority belongs.—Ed.

substances which were the results of a secondary chemical and not strictly electrical action on the substances exposed to the influence of the electrical curre nt.

Iron. — Oxide of.
„ Mammillated black oxide of.
„ Sulphuret of ÷
Copper. — Mammillated.
„ Phosphate.
„ Grey sulphuret.
„ Dendritic.
„ Green and blue carbonate.
„ Chrysocolla.
„ Copper pyrites.
Copper. — *Sub*sulphate of (new mineral).
Lead. — Dendritic.
„ Yellow oxide of.
Tin. — Dendritic.
Gold. — Dendritic.
Silver. — Dendritic.
„ Capillary.
„ Oxide of ÷
„ Sulphuret of.
Antimony. — Sulphuret of.
Silex. — Dendritic.
Alumina. — In nodules.
Lime. — Carbonate of ÷
Strontia. — Sulphur.
Chalcedony.

Mr. Crosse made some very interesting and important experiments in connection with the separation of metals from their ores. The following are details of experimental arrangements that he was in process of conducting at the time of his disease.

"*On an entirely new Mode of extracting Gold from its Ores.*

"I took a piece of quartzy gold ore from California, which weighed 4306 grains, and reduced it to a coarse powder in an iron mortar. This I exposed to a dull red heat for one hour and a quarter, in a black lead crucible; after which its weight was 4260 grains, having lost 46 grains by the roasting. I then reduced this coarse powder to a fine one, and carefully and intimately mixed the whole together, so that the gold it contained should be as equally disseminated throughout the whole mass as possible.

"*Experiment* 1*st.*—Of this powder I took 1000 grains, and put them into a Wedgwood mortar, having first thrown into it 200 grains of pure mercury. I then partly filled the mortar with extremely dilute carbonate of ammonia, and connected the mercury with the *negative* pole of a VERY WEAK voltaic battery of twelve pairs of cylinders, keeping up the action for five hours; after which I disconnected it with the battery, and stirred up the mixture for some time, grinding it with a Wedgwood pestle. I should observe that, after I first connected the mercury with the negative pole of the battery, — which I did by a wire of platinum, — I connected, by a similar platinum wire, the supernatant fluid with the positive pole, taking care not to immerse the end in the fluid more than about one inch and a quarter deep, and keeping it near the edge of the mortar, so as to prevent the mercury, when expanding, from coming into contact with the positive wire, which would of course destroy the effect of the electric action.

"After I had finished grinding the mixture I decanted the fluid, and washed the residuum with repeated effusions of cold water, and carefully poured off the mercury, which was decidedly more fluid and clearer than it was when first used. The earthy residuum I divided into two parts, throwing the lighter portion into one vessel, which I call A, and the heavier, which I call B, into another vessel. A consisted principally of peroxide of iron, and had lost some sulphur in the roasting. It also contained a certain portion of silica, and was of a reddish ochrey colour. B was composed almost entirely of quartz, which was remarkably white and clean.

"I next weighed the mercury, having carefully dried it. Its weight was 205 grains, which, when evaporated in a black lead crucible yielded *eight* grains of gold.

"My next object was to discover, by chemical analysis, if any, and what, gold had been left untouched by the above process in the powders A and B; and accordingly I digested A in nitro-hydrochloric acid for three hours, and then tested the solution by adding sulphate of iron, which did not occasion the slightest precipitate, nor was there a trace of gold to be found in it.

"I then repeated the same process with B, which, being less in quantity, I digested only for two hours and a quarter, and by applying the same test I obtained a precipitate of pure gold, which, when dried and weighed, amounted to *half a grain;* so that the above electrical process had only left, after five hours' action, one seventeenth part of the gold untouched.

"*Experiment 2nd.*—I next compared the common mode of extracting the gold by amalgamation with that first described. I took 1000 grains of the same ore as the last. This I threw into an iron mortar with a quarter of a pint of common cold water, and poured into it 200 grains of mercury, stirring it up with an iron pestle for a considerable time; after which I allowed it to rest five hours, and then I again stirred it up well for some time. I next decanted the water and washed the residuum by repeated affusions of cold water; and with much trouble I separated the mercury from the mass, which did not run off from it so freely and easily as in the former experiment. I then divided the residuum, as before, into two portions, the lighter of which I call A and the heavier B. Each of these portions much resembled those obtained in the first experiment. I next dried the mercury and weighed it. Its weight was 197 grains, which, when evaporated in a black lead crucible, yielded barely four grains of gold.

"I then digested A for two hours and a half in nitro-hydrochloric acid, and tested it for gold, but not the least sign of gold was visible.

"I next digested B for the same time in similar acid, and by adding solution of sulphate of iron I threw down *four grains of gold.*

"*Advantages of the Electrical Process.*

"The swelling of the mercury is caused by the formation of the well-known ammoniacal amalgam, and which rises so high as to embrace the lighter particles of the ore which float on the surface, and from which the precious metals are more readily extracted. Likewise the mercury at the bottom is rendered *far more attractive* of gold and silver than it would be in its natural state; as a proof of which the negative platinum wire is immediately covered with mercury, though in its unelectrified state it has not the least affinity for that metal. Again, the mercury is kept by the negative electrical action in a far greater state of fluidity and comparative purity than it is in the common mode of amalgamation, and is consequently much better adapted to separate the valuable metals. Lastly, a much less quantity of mercury is required than in the usual mode. A comparatively weak voltaic battery will suffice to keep a large body of mercury in a state to separate a large quantity of the precious metal."

FURTHER EXPERIMENTS.

"April 10th, 1855. Set in action a battery of twelve large double pairs in constant action, upon a large basin of dilute solution of salt and nitre, with 1000 grains of Californian gold ore, in fine powder, thrown upon a plate of platinum, made *positive*, and suspending a small glass cup with platinum coil within it, made *negative*. In twenty-four hours I took up the glass cup, and found that it contained rather more than one grain of pure gold. I replaced the glass cup and renewed the action.

"April 17th. Took out the glass cup again and found I had obtained five grains of pure gold, making in all *fully* six grains of pure gold and two grains of silver; total, eight grains, being the whole proportion of metal contained in the 1000 grains of ore. N. B. A good experiment, but too slow for use."

In another experiment Mr. Crosse sent an electric current through a dilute solution of salt and nitre, acting upon two half sovereigns. In a few days it was found that they had lost thirty-one grains, which was deposited at the negative pole of the battery.

Separation of Copper from its Ores by Electricity.

The result of Mr. Crosse's trials upon the separation of copper from its ores was, scientifically speaking, perfectly successful; and the only drawback to the more extended application of the process appears to be the expense of the battery. Nothing can be more beautifully simple or more complete in its results than the arrangement. For all purposes of analysis it is most accurate. The plan is as follows (the description refers to a comparatively small experiment, where a few pounds only of ore are used): —

At the bottom of a large glass (or wooden vessel) a piece of platinum gauze is let fall, on which is thrown a certain proportion of calcined copper ore in powder. This platinum gauze is connected with the positive pole of the battery by a platinum wire covered with gutta percha. A small glass or porous cup is suspended by platinum wires, and let down into the larger vessel, but kept several inches above the ore at the bottom. A coil also of platinum wire connected with the negative pole is dropped into the cup. The large vessel is then filled with dilute sulphuric acid, viz., about forty-nine parts of water to one of acid (for purposes of analysis the proportion of acid may be greater to facilitate the result).

The electric circuit being complete, the acid is concentrated at the positive pole, and acts on the ore; the copper contained in it is shortly transferred to the negative coil, within the receiving cup, and is there found deposited in a perfectly pure metallic powder. If the experiment is properly conducted the result will be that the whole of the copper is deposited in the cup, from whence it can be removed, dried, and weighed,—giving the per-centage of the ore, or, if conducted on a large scale, supplying chemically pure copper by a process at once simple and efficacious.

This mode of extracting metals from their ores has been patented; it remains yet to be seen whether the expense of the battery admits of its general application to commerce. Mr. Crosse used to say that if he could construct "a battery at once cheap, powerful, and durable, he might say with Archimedes that he could move the world."

The following note appears relative to the kind of battery useful for the electrical reduction of ores:—

"A battery which I think," says Mr. Crosse, "most available for the extraction of copper from its ores, consists of a certain number of earthen jars (or glass), in each of which a *porous pot* is placed, commensurate with the size of the jar. In each jar is placed vertically a cylinder of thin sheet copper, which surrounds the porous pot that stands in the centre. In each porous pot a rod or cylinder of stout zinc is placed vertically, the top of which rod is connected by means of a strip of sheet copper with the top of the adjoining copper cylinder. In order to excite this battery, each zinc rod is first covered or amalgamated with a thin coating of mercury. Next each porous pot is filled with dilute sulphuric acid.

consisting of one part of acid and nineteen parts of water. Lastly, the earthen or glass jar is filled with dilute sulphuric acid of the same strength with the above. But in the last must be previously dissolved as much powdered sulphate of copper as the fluid will hold in solution; moreover, a certain quantity of dry powdered sulphate of copper is thrown into the latter solution, to feed or sustain the action of the battery. This battery is called Daniel's Sustaining Battery; but, for the purpose to which I apply it, it is used with a *much* less proportion of acid than commonly. The principal expense of this battery is the consumption of the zinc and the acid (the copper being little acted on), and also the consumption of the sulphate of copper. As a set-off against the expense of using this battery, the zinc and copper in solution are both recoverable."

More than five and twenty years ago Mr. Crosse gave it as his opinion that the time would come when metals would be separated from their ores by electricity.

Mr. Crosse directed his attention to the purification of sea water and other fluids by electricity. In acting on sea water, it is necessary to submit it to one distillation, a process which only renders it fit for washing, but not for drinking. A very simple electrical arrangement is then applied to the cask or cistern of sea water; and in twenty-four hours or less the water becomes perfectly wholesome to the taste, and will remain sweet in open vessels for an indefinite time. The mode of electrisation is curiously simple. Two cylinders, of dissimilar metals (generally sheet zinc and sheet iron), are placed in two porous earthenware tubes, open at the top, and closed at the bottom. The metallic cylinders being connected together by a copper riband, the porous tubes, with the metals inverted in them, are filled with water and placed in the fluid required to be purified. The electrical action immediately commences, and the fluid not only becomes purified, but is *rendered* ANTISEPTIC in a few hours.

"The application of this principle to wines and brandy, has been attended with great success. It has the effect of softening the asperities of some wines by removing the predominance of bitartrate of potash; and in the case of the spirit distilled in imitation of French brandy, the improvement to be derived from using the process is remarkable." In one experiment, two gallons of the very worst English brandy were kept electrified for three weeks: at the end of the time the spirit was drawn off infinitely improved, indeed *visibly* purified, for the water in the positive porous tube had become intensely acid, and the negative tube was filled with a green oleaginous fluid, thick and turbid.

This process has also been applied most effectually to stopping the *fermentation* in cider, and also in other things, such as starch. Brackish water is also wonderfully purified by this arrangement.

(This application of the electric principle has also been patented by a company of gentlemen.)

The antiseptic power of electrified water is very remarkable. Not only can it be preserved for years perfectly clear and fresh, but it has the power of restoring the most putrid substances to sweetness. Pieces of meat and the skins of animals in a state of putridity have been immersed in electrified water, and in a few hours rendered inodorous. Milk has also been kept sweet for three weeks in the middle of summer, by the application of electricity. On one occasion Mr. Crosse kept a pair of soles under the electric action for three months, and at the end of that time they were sent to a friend, whose domestics knew nothing of the experiment. Before the cook dressed them her master asked her whether she thought they were fresh, as he had some doubts. She replied that she was sure they were fresh ; indeed she said she could swear that they were alive yesterday ! ! When served at table they appeared like ordinary fish; but when the family attempted to eat them they were found to be perfectly tasteless: the electric action had taken away all the essential oil, leaving the fish unfit for food. However, the process is exceedingly useful for keeping fish, meat, &c., fresh and *good*, for ten days or a fortnight. I have never heard a satisfactory explanation of the cause of the antiseptic power communicated to water by the passage of the electric current. Whether ozone has not something to do with it, may be a question. The same effect is produced, whichever two dissimilar metals are used. It often occurred to Mr. Crosse that the electrified water might be drunk beneficially in cases of typhus and other fevers, and also could be used for baths.

The subjoined are extracts from a paper read by Andrew Crosse at one of the meetings of the Electrical Society, of which he was a member. The subject is,—

"*On the Perforation of Non-conducting Substances by the Mechanical Action of the Electric Fluid.*

"I find that if an intercepted current of electricity from a common electrical machine were passed between points of two platinum wires well secured on a narrow strip of common window glass, — both wires being on the same side of the glass, forming a straight line, the two nearest ends of each being at the distance of the tenth or twelfth of an inch apart, and the two more distant ends being respectively connected with the positive and negative conductors of the machine, the glass strip with its wires being let fall into a vessel of water, — after turning the handle of the machine 200 or 300 times, a small, perfectly circular, and tolerably even hole was drilled in the glass, between the two wires, and at *right angles* to the direction of the electric current. * * I repeated this experiment in a variety of ways. In the arrangement, that side on which the wires are fixed is uppermost, and covered with water to the depth of an inch; but when the water barely covers the glass in this situation, on turning the machine, the water immediately above the interval between the wires is struck up to some height, upwards of a foot from its surface, continually playing like a minute fountain. If on the other hand the glass strip is inverted and placed with the wires on its lower surface, with the water scarcely covering its upper surface, on turning the machine very minute pieces of glass are displaced from that part of the strip immediately above the space between the two wires, and a small jet of particles of glass is constantly thrown up, as the water was in the former instance. * * I fixed two wires at the distance of $\frac{1}{20}$ of an inch on a strip of good plate glass, $\frac{1}{8}$ of an inch thick, under water. After 500 revolutions it was pierced to the depth of about $\frac{1}{3}$ of its thickness, but on the same side on which the wires were fixed. The other side was not in the least degree affected. When the glass is thin a series of vibrations is communicated to it, sufficiently powerful to separate a small portion of its substance from the *opposite* side; but when plate glass is used this effect does not take place — apparently on account of its not being susceptible of being thrown into similar vibrations; and as it presents a stronger resistance, the effect is produced on the same side on which the wires are fixed. I next fixed the wires very firmly on the side of a perfectly transparent crystal of quartz, at a distance of $\frac{1}{20}$ of an inch. After 100 revolutions, a small mark was visible between the wires, but the surface of the crystal remained smooth. I then immersed it in a glass dish of water, and passed 100 revolutions across it.

On examination I found a very sensible excavation between the wires, on the same side of the crystal. After 500 revolutions it was pierced to a much greater depth; this increased effect was produced by the wires having been more firmly secured on the surface of the crystal. I am inclined to think that, with proper apparatus, well contrived to secure a firm pressure of both wires on the substance experimented on, and by proper regulation of the distance of the extremities of the wires, together with a sufficiently powerful stream of electricity, two or three times interrupted, not only glass may be sufficiently well and readily drilled, but even the diamond and other gems may be perforated in the same way."

I understand that a modification of this plan has been adopted successfully by medical practitioners in cases of internal disorders, which are very difficult to treat with ordinary means. A paper was read before the Royal Society, some four or five years ago, on this subject, in which reference was made to Mr. Crosse's first experiment on the perforation and breaking to pieces of non-conducting substances by electricity.

Electro-vegetation.

One of the most interesting theories relative to electricity is that which treats of the connection between the growth of vegetation and the electric influence. It was Mr. Crosse's decided opinion that electric action (possibly local electric action) was the cause of mineral substances being carried into, and forming component parts of, the vegetable kingdom, and would account for so insoluble a substance as silica forming so large an element in vegetable productions. Mr. Crosse says: "A field of boundless extent lies open to reward him who may closely follow up these researches, although the best mode of conducting such experiments remains to be discovered."

Professor Quekett has enlarged upon and adopted

this view in his "Lectures on Histology,"* lately published.

"When we consider the vast amount of silica that must be taken from the soil by the straw of grasses of various kinds, it is possible that, besides the nitrogenous principle which guano contains, the silica in it may also be of considerable service. It is certain that the cereal plants must take it up from the soil, for the atmosphere cannot supply it, and it could hardly be given to them in a more finely divided state; thus constituting another valuable quality of this material as a manure. The process of dissolving the silica and taking it up to be deposited in the tissues, as is done by the grasses, is probably an electrical one; and in a recent visit to Somersetshire I witnessed the following most striking experiment in the laboratory of Mr. Crosse, a true philosopher, of whom doubtless you have heard as being so celebrated for his experiments in voltaic electricity. In a common tumbler filled with *distilled* water, were placed, on opposite sides, a portion of silver—if I recollect rightly, a sixpence †— and a piece of slate; one was connected with the positive and the other with the negative pole of a voltaic battery, consisting of a vessel containing a solution of sulphate of copper, in which was placed a porous tube and a zinc rod, the tube being full of a solution of sugar: by this means slow electric action was

* Lectures on Histology, by John Quekett, vol. ii. p. 68.

† It was chemically pure silver.—Ed.

kept up, and the silver on the one side was actually dissolved by the water, carried across, and deposited in a crystalline form upon the slate on the opposite side of the tumbler. Had a piece of flint occupied the place of the silver, the same effect in all probability would have been produced. It occurred to me immediately that it might be by an electrical agency that the silica, lime, and other inorganic materials were dissolved and assimilated by plants."

Mr. Crosse tried the following experiment upon the potatoe. He says: —

"I took two garden pots; I stopped up the holes in each with a cork, and then, filling them with earth, placed them side by side in close contact with each other in a large pan of water. I next planted a single potatoe in each pot, and made a conducting communication by means of two platina wires between the earth in each pot and the opposite poles of a voltaic battery, in weak but constant action. One end of each wire was plunged about four inches deep into the earth of each pot, but at a distance from and not touching the potatoe. Thus, one of these potatoes was planted in positively electrified earth and the other in earth negatively electrified. After a while the negative potatoe contracted the disease, was decomposed, emitted the peculiar fetid smell of the diseased potatoe;—the garden-pot was filled with the same kind of insects which infest the diseased plant. On the other hand the positive potatoe did not contract the disease, nor did it emit any smell nor yield any fetid liquor, nor was there a single insect visible within the positive pot: the effect, however, on the positive potatoe was most singular, for, when removed from the earth, it appeared that it had neither shot out root nor stem, but, whilst perfectly solid and unbroken, it extremely resembled a shrivelled apple, both in smell and appearance."

I have often heard Mr. Crosse observe upon the following curious circumstance. He invariably found that *negative* electricity was injurious to all vegetation, except the development of fungi. *Positive* electricity, on the other hand, he found most favourable to all vegetation, except all fungoid appearances, which it entirely checked. In the course of his experiments he constantly found fungi growing in copper and other acid solutions. On one occasion a mushroom-shaped fungus grew out of electrified hydrosulphuret of potash; and frequently I have myself seen the surface of an electrified fluid covered, or nearly so, by a thick flesh-like fungus that was strong enough to bear a considerable weight, and which was so tough as hardly to be torn apart by the fingers.

Mr. Crosse considered that the roots and leaves of plants were in opposite states of electricity, and he often proposed trying to make a battery of growing plants, or at least an arrangement that might prove that electricity was present. I remember his description of a very elegant experiment on some roses. He had two branches cut from the same tree; they were as nearly alike as it was possible, with the same number of buds, and both equally blown. An arrangement was made by which a negative current of electricity was passed through one, a positive current through the other. In a few hours the negative rose drooped and died; but the positive continued its freshness for nearly a fortnight, the rose itself became full blown, and the buds expanded and survived an unusual length of time.

Mr. Crosse was not very sanguine about artificial electro-vegetation, from the great and almost insuperable difficulty of applying electricity to any extent. In the case of single plants it may be easy enough. Mr. Crosse himself says, in a letter to John Boys, Esq.: —

"Broomfield, near Taunton,
"February 1st, 1847.

"Sir,

* * * "It is very true that I have during a long period of many years given a large portion of my time to the investigation of the science of electricity, but I am not

partial to drawing hasty conclusions, as *cause* and *effect* are so extremely difficult to distinguish from each other, in this as well as other matters, that the deepest thinkers have often mistaken the one for the other. The qualities of *light, heat, magnetism,* and *electricity,* are thus co-existent and intimately blended together, and some of our first philosophers have asserted that they are one and the same; but most assuredly they have not proved it scientifically. As to atmospheric electricity, it occurs invariably, to a greater or less degree, after a more or less rapid evaporation of the moisture of the earth's surface. Thus, in those countries where a succession of rain and solar heat alternate most frequently and in the greatest intensity, thunder-storms are the most numerous and violent; and therefore, in this case, electricity may be ascribed to the evaporation of water. Again, as magnetism and electricity always accompany each other, the current of the one being at right angles to that of the other, the globe, which is polarised magnetically, must be belted by electrical currents, affecting, most probably, the interior of the earth more than its atmosphere, the earth being a good conducting medium and the air being a *non*-conductor of electricity. To what extent electricity may act upon the atmosphere above, or on the earth below; as an *exciter* of heat, I am unable to say, but in this respect it appears to me rather the *consequence* than the *cause.* The heat precedes the formation of thunder clouds, and the temperature becomes much lowered by the discharge. I have no doubt in my own mind but that electrical action is essential to vegetation; but little or nothing definite has been done in that department, in which men of science have taken very opposite views. For myself, I in general find that *positive* electricity is a great promoter, and that *negative* electricity is a great checker, of vegetation, but this mostly in consequence of the substances in which the plant is growing being more or less favourable or adverse to its growth, being electrically attracted, or diverted *to* or *from* the roots of the plant acted on. This is best effected by plunging the plant into a pot of moist earth, which pot is of a porous material as usual, but with no hole at the bottom. This pot is to be placed in a basin of water, the level of which may be about an inch below the edge of the pot. A wire of platina should be stuck into the moist earth of the pot, a few inches deep, the other end being connected with the positive pole of a small galvanic battery; also a similar platinum wire must connect the negative pole of the same battery with the water in the basin. No insulation is necessary. If I were to electrify the roots of a vine growing in the common ground, I should provide a copper wire with a termination of platinum wire, which termination I should stick between the roots of the vine about twenty inches deep, connecting the copper end with the positive pole of a small battery under an adjacent shed, and making use of a simple copper wire to connect the earth with the negative pole,—such wire being about twenty inches deep into the ground, at the distance of about four feet or two yards from the platinum wire. There would be a mutual interchange of elements; and such vine might be compared with a vine under the same circumstances, but not electrified. The electrified vine should be kept well watered, to form a good conducting communication between the two wires, and the unelectrified vine should be equally watered, so as to form a fair test. The battery made use of should be Daniel's Sustaining Battery, about eight pairs of plates excited by sulphate of copper in the negative cells, and water alone in the positive.

"I beg to remain, sir,

"Your obedient servant,

"ANDREW CROSSE."

I venture to give Mr. Boys' letters in answer, as they contain the results of experiments tried at the suggestion of Mr. Crosse: —

"Margate, 29th July, 1847.

"Dear Sir,

* * * * * *

"I obtained in March last two vines raised from eyes of the same shoots, one year old. I separated the roots of each, and cut away all fibres, leaving only six roots to each plant; and those roots I also cut back, to the same length for each. I procured two flower-pots for the two vines each 18 inches deep, and 16 inches across. I stopped the holes at bottom, and planted each vine in precisely the same mixture of soil, as well as contents of each pot.

"I placed each pot in a large crock, so that I surrounded each internal pot with rain water, and kept them constantly filled; the copper wire was plunged from the positive pole of the battery to the roots of one vine, and the negative wire was placed in the water outside the pot. About twice a week each plant was watered with about half a pint of rain water; and the situation in which I placed the two in reference to temperature, light, and air, was with equal care observed.

"A small galvanic battery, as advised by you, was then applied to one vine, the other being perfectly detached and left to the operation of unaided nature.

"I commenced on the 30th March and finished on the 30th June.

"The result was as follows: —

"The electrified vine shot 74 inches.

"The unelectrified vine shot $31\frac{1}{4}$ inches.

"Each bore a bunch of white Dutch grapes, the electrified bunch being about $\frac{1}{3}$ larger than the unelectrified bunch.

The electrified stem is the largest in girth and it has (on this day) twenty-nine *ripened* joints, whilst the unelectrified has only nine, and in all other appearances the electrified is decidedly the best plant.

"I propose next season to repeat exactly the same process in every point, except that I will apply the galvanic process to the hitherto unelectrified vine, in order to discover if possible whether there might have been any difference in the original healthiness of the two plants.

* * * * * *

"I remain, &c. &c.

"JOHN BOYS.

"Andrew Crosse, Esq."

The following season Mr. Boys writes Mr. Crosse, giving him the result of the second year's experiment, in form of an abstract from his diary: —

"At the suggestion of Mr. Crosse I repeated the experiment in 1848 upon the same two vines by placing No. 2. vine under the same galvanic process as No. 1. underwent in 1847, and I continued the experiment for three months from the 22nd February to the 22nd May, 1848, being the same length of time as it was applied in the preceding season. The result was as follows: —

"No. 1., which under the galvanic process in 1847 had grown 74 inches, did in 1848, ungalvanised, grow only 68 inches, thereby growing less by 6 inches; whereas, No. 2., which in 1847, ungalvanised, had grown 31 inches, did in 1848, under galvanism, grow 60 inches and $\frac{1}{8}$, thereby *growing more* by 30 inches than it had done in 1847, and nearly *doubling* that prior year's growth.

"To the above statement I think I shall have to add a further proof when the two vines have completed their present season's growth; for during the six days that the battery has been withdrawn each vine has grown at an equal rate.

* * * * * *

"I remain, &c. &c.,

"John Boys."

Discovery of a new Mode for making Impressions on Marble.

"In a saucer filled," says Mr. Crosse, "with a concentrated solution of nitrate of potassa, I placed horizontally a flat polished piece of white marble, and upon the middle of the marble a common sovereign, with its reverse in contact with the marble, and having a stout glass rod supported perpendicularly on the coin to keep it in its place. Between the rod and the coin was affixed a platinum wire, which was connected with the positive pole of a sulphate of copper battery of eight pairs of plates; while round the marble, but not touching it, was a coil of similar wire, connected with the negative pole. The nitric acid was soon separated from the potassa, and attacked the marble in contact with the sovereign, and at the expiration of three days the coin was perfectly embedded in the marble. The experiment was then stopped, and the marble being taken out and inverted, the sovereign fell out of its stony receptacle, leaving a very perfect impression on the marble, the surface of which still retained its polish. There is one difficulty attending this manner of taking casts, viz., that the close contact between the two surfaces of the coin and the marble does not allow of a sufficiently easy access to the decomposing fluid, and consequent escape of the carbonic acid gas. Casts therefore formed in this way are necessarily deeper at the edges, where the stone is first acted on; but if impressions were formed by wires woven into a sort of gauze-work of gold on platinum, the effect would be perfect."

Mr. Crosse found that by sending the voltaic current through common plaster of Paris, it became *indurated* and *crystallised* throughout. The external surfaces looked as if polished, and the induration was so complete that the plaster of Paris became as hard as the hardest marble. This effect was produced by a single arc of zinc and copper.

The following memoranda appear: —

"In the course of some experiments in which I had occasion to use dilute nitric acid acting on copper ore in the electric circuit, I found after two days' action that the acid was *decomposed*, its oxygen having passed to the positive pole, and a small porous pot, which was made negative, was filled with *aqueous ammonia*. Thus nitric acid, after electrisation, was converted into ammonia."

Mr. Crosse performed repeated experiments, in

which he dissolved chemically pure gold and silver in *distilled* water, with a battery of two pairs of plates weakly charged. In a letter to Mr. Weeks, he says: —

"I have succeeded in dissolving *largely*, *pure* silver in distilled water, by electric action on a solid mass of it."

The metal, according to the strength of the electric action, was transferred from the positive pole —either in powder, and deposited at the negative, or, with a slower action, it became arranged at the negative terminal in a crystalline form.

Blood kept Fluid.

Two ounces of pig's blood was kept in a state of electrisation from February 1854 to February 1856 (the latter date subsequent to Mr. Crosse's decease). It remained perfectly fluid to the last; the colour was unaltered. It was acted on by three pairs of plates, — sulphate of copper battery, which was occasionally fed to maintain the action. Whether electricity may not have the effect of retaining the ammonia which, according to the paper read by Dr. Richardson at the Cheltenham meeting of the British Association, appears necessary to preserve the fluidity of the blood, is only an idea suggested by the experiment; but the question seems an interesting one.

Analysis of the Voltaic Battery.

"1st. I find that a disc of zinc being let fall into water, *distilled*, *common*, or *acidified*, being removed with an insulating handle and carefully examined, is in a *positive* state.

"2nd. I find that the fluid from which it has been removed is in a *negative* state.

"3rd. That a disc of platinum being let fall into similar water, and similarly removed, is *not electrified*, and the fluid from which it has been removed is *not electrified.*

"4thly. If a strip of sheet zinc be *bent* over the edge of the basin containing the above fluid, in such manner as that one part of the zinc is plunged into the fluid, and the other half is kept outside and dry, while the wet part is *positive*, the *dry part is negative.*

"5thly. If this experiment be repeated with a strip of platinum alone, there is no electricity either within or without the basin.

"6thly. If both strips, viz., the zinc and platinum, are plunged into the fluid, their wet sides opposing but not touching each other, the wet part of the zinc is *positive* and its dry outside *negative*, as before; but in this case the wet part of the platinum is *negative*, whilst its dry outside is *positive*: however, no current is produced, till the outside of each be brought into contact or conducting communication with each other. The current will then be as follows. It commences with the wet part of the *zinc alone*, from which issues a *positive* current, electrifying the platinum opposed to it *negatively* by induction; the *positive* current from the zinc passing through the *negative* electricity of the platinum, forming, when *combined* with the outside of the zinc, a regular voltaic circle, through which a constant electrical current passes. That the zinc or most oxidable metal is the *original motive power*, which sets free a portion of the natural electricity of the *water*, which natural electricity is employed in keeping together the due proportion of oxygen and hydrogen, necessary to form water. The zinc being, as Davy calls it, *electro-positive* and the oxygen *electro-negative*, the oxygen passes to the zinc, while its corresponding hydrogen is set free, if zinc alone be employed; but if platinum be opposed, the hydrogen passes to the platinum. In either case the water is decomposed, and the electricity which was previously locked up is liberated, and goes to the zinc in an accumulated state, but not producing a current, without employing an opposite, less oxidable metal, the outer termination of each metal being connected together by a conducting medium, through which the electric current passes. The water thus furnishes the materials for its self-decomposition by the agency of the zinc, and the platinum being merely a conductor. This does away with the contact theory, and also with some others."

Experiment called the Mine in a Garden Pot.

"A A, a wide large basin filled with water to the level B B. C C, a wide *porous* pot, plunged into the basin of

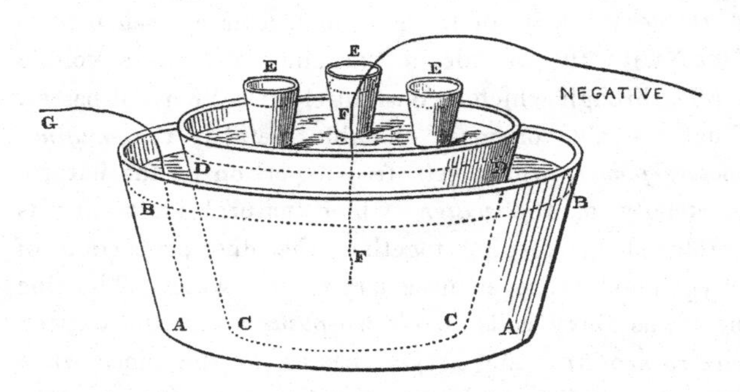

water, and filled to the level D D with pipe-clay kneaded up with water to the consistence of moist putty. E E E, three funnels plunged vertically into the moist clay of the porous pot, and standing so as to form an equilateral triangle with each other. One of these funnels was filled with a solution of nitrate of silver, the second with sulphate of copper, and the third with hydrosulphuret of potash.

"A platina wire, F, was plunged vertically into the clay, about one half way into it, and in the middle of the porous pot, between the three funnels, at an equal distance from each. This wire F was connected with the negative pole of a *small* constant voltaic battery. G is another platina wire plunged into the water of the basin A A. This wire I coiled two or three times round the outside of the porous pot C C, the wire being connected with the positive pole of the same battery. It is evident by this arrangement the electric current enters the water G G, passes through the porous pot C C, and the moist pipe-clay therein, and goes out by the wire F to the other pole of the battery. Consequently the water in the basin A A is *positive*, and the porous pot C C is *negative.* The effects were as follows : —

"1st. In the course of a few days two transverse fissures took place in the moist clay at F, which fissures extended downwards perpendicularly till they reached the bottom of the pot.

"2nd. Immediately afterwards the water from the external basin, being attracted through the porous pot, rose to the top of the fissures, and formed a little pond in the middle of the clay, at length filled up the top of the porous pot C C, and finally trickled over the lowest part into the basin below.

"3rd. In the course of some months, the different solutions in the three funnels E E E slowly percolated the clay, by mechanical pressure, gradually mixed together, and were slowly decomposed by the electrical agency. The fissures being filled with water, and presenting the best conducting medium, formed a reservoir for the mixed fluids, which separated according to their electrical affinity. The new compounds arising from which formed crystalline and amorphous substances which lined the fissures, and converted them into *metallic lodes.* These substances consisted of metallic silver, sulphuret of silver, metallic copper, red oxide of copper, sulphuret of copper, and crystallised sulphur.

"I kept up the action for between one and two years, and then took the whole apart, and found what I have described. First, we have the fissure occasioned by negative electricity. Secondly, we have the spring, or rather water raised above its level, by negative electricity. (In this way I have formed a perpetual spring, and raised some ounces of water several inches above its level in the twenty-four hours.) All this is done by a weak voltaic battery. It may be objected that the water is drawn into the fissures by capillary attraction. It is easy to prove that this is not the case. Reverse the passage of electricity, let it enter at F and go out at G, make F positive and G negative; none of these effects will take place, but the water previously mixed with the pipe-clay to moisten it, will be drawn out of the clay, and added to that in the basin A A.

"In order to cause an earthquake in the pot of clay, make it negative as at first, leave out the use of the funnels, and connect it with a rather powerful battery: within twenty-four hours the moist clay surrounding the negative wire F will be thrown up in all directions to the extent of several cubic inches. Here we have the earthquake."

At the chemical section of the British Association Meeting in 1854, Mr. Crosse made the following communication:—

" *On the apparent Mechanical Action accompanying Electric Transfer.*

" *Experiment* 1*st.*—I placed a piece of smooth carbonate of lime, of two inches square, and half an inch thick, at the bottom of a rather deep saucer, which I nearly filled with dilute pure nitric acid. The preparation of the acid being $\frac{1}{50}$ part by measure of the distilled water employed, which was one pint. Upon this piece of limestone I placed a sovereign, which weighed 123 grains, and upon the upper surface of the coin I placed one end of a platinum wire, which was connected with the *positive* pole of a small sustaining sulphate of copper battery. This end of the wire was kept firm on the coin, and the coin on the limestone by a stick of glass, supported vertically. The lower end of the stick was ground. Around the square piece of limestone I coiled a second platinum wire, which was connected with the negative pole of the battery.

" The action commenced, hydrogen gas being liberated at the latter pole, and carbonic acid gas from that part of the coin in contact with the limestone at the positive pole. I kept this in action for fifty hours, and then took the apparatus apart. The coin had sunk into the limestone to the depth of half its thickness, and when removed it left a clear impression on the stone. But the most striking circumstance was that the carbonic acid gas, in its evolution from the stone, had struck off a portion of the milled edge of the sovereign, leaving it quite smooth at that part, and the pieces *broken* off had the milled edge remaining on them. Moreover, the evolution of gas, carried up a small portion of gold, and gilded the whole of the ground surface of the glass rod. The broken pieces of metal lay around the coin, which when weighed showed a loss of three grains, which was exactly the weight of the pieces, including the gilding on the glass, which I carefully removed. It is particularly to be noticed that this was at the *positive* pole. On testing the fluid it evinced not a trace of *gold* or *copper*, but merely a portion of nitrate of lime. Indeed had either of these metals been in solution they must have appeared on the *negative* platinum wire; which was not the case.

" *Experiment 2nd.*—I repeated the former experiment in a different manner, using pure sulphuric acid instead of nitric, and acting on the same sovereign, which now weighed 120 grains. This I placed on a larger piece of marble, and kept it pressed firm in its position by a glass weight of larger dimensions than the former, and weighing about two pounds. Instead of a saucer I used a glass jar, filling it with one ounce of sulphuric acid, and forty-nine of distilled water ; so that the pressure of the fluid was of course greater, from the greater depth of the vessel containing it, and resistance to the extrication of the gas was in consequence proportionally greater. I employed a sulphate of copper battery of eight pairs, weakly charged, but in good action. This action was continued for ninety hours, and then stopped. The coin weighed 114 grains, having lost six grains, which lay in pieces around it upon the surface of the marble, and which weighed exactly six grains. The glass weight in this experiment was not gilded, and the coin had made but little impression on the marble. On examining the sovereign, I found that one portion of its edge had the entire milling completely removed, and that part of its edge was left perfectly smooth, the remaining part of the coin being little, if at all acted on. In fact, neither of the flat sides of the coin was at all acted on: with the exception of *both* sides which were contiguous to that part of the edge from which the milling was removed. The carbonic acid gas, which was liberated from the limestone, had found an easier vent from under one part of the coin than the other, and from this part it poured forth in considerable quantity, and by its *constant friction* broke off small pieces of the coin which lay in a heap adjoining. I must observe that a minute quantity of the purple oxide of gold stained a part of the marble."

Three days before an attack of illness which ended fatally with Mr. Crosse, he tried a further experiment on this subject, the result of which was noted down by myself, and communicated to the public through the same channel as the notice of the first experiments.

"*On the apparent Mechanical Action accompanying Electric Transfer.**

"The experiments detailed by the late Mr. Crosse on

* The Notes of the Experiments, by Mrs. Andrew Crosse, read by the President of the Chemical Section, Dr. Lyon Playfair, at the Meeting of the British Association at Glasgow, in 1855.

this subject at the Chemical Section of the British Association Meeting at Liverpool, were briefly these."

(It is unnecessary here to repeat the account.)

"Objections were raised to these experiments. It was stated that the same results would not have occurred if *pure* gold was made use of, instead of a sovereign. It was suggested, that the nitric and sulphuric acids acted on the alloy of the sovereign, and, in fact, that the removal of the edges of the coin was effected not by the mechanical action of the gas as stated by Mr. Crosse, but by the disintegration of the copper; and that the purple stain that was perceptible on the limestone, was attributed to the presence of copper.

"In the early part of May, 1855, Mr. Crosse procured from Messrs. Johnson and Matthey, assayers, Hatton Garden, a piece of *chemically pure gold*, prepared most carefully and expressly for the experiment. This piece of solid gold was about the size and shape of a florin, and weighed 201 grains.

"It was placed on a piece of pure white carbonate of lime, or marble, about four inches square. A ball of glass weighing about two pounds was placed on the gold, the object of which was to keep in its place the end of a platinum wire, which was placed immediately under the gold in connection with the limestone. This platinum wire was the terminal of the positive pole of a sustaining sulphate of copper battery of twelve pairs, strongly charged. The limestone supporting the gold and the glass ball was placed at the bottom of a twelve-inch glass jar, which was filled up with pure sulphuric acid and distilled water ; the proportions being forty-nine of water to one of acid. An end of platinum wire was coiled round the limestone connecting it with the negative pole of the battery. When the circuit was completed a rapid action commenced, the gases were violently liberated at their respective poles. In twelve hours the apparent mechanical action of the carbonic acid gas, at that side where the extrication was the most free, had broken off pieces of the gold, which lay around on the surface of the limestone.

" At the end of twenty-four hours the collection of bits of gold was very considerable. In about four days the action of the battery became visibly weakened, and the deposition of the pieces of the gold considerably lessened.

" Looking edgeways at the florin-shaped piece of gold, I observed that it had the appearance of being separated into collateral flakes; and I found, on taking apart the experiment, that such was the case to a considerable extent. Indeed, one large piece had become separated or scaled off from the solid mass, which measured an inch and a quarter on the outer circle, and was half an inch in breadth, and of about the thickness of writing paper.

* * * * * " A stain of the *purple oxide* of *gold* was perceptible, as in the former experiments, fully proving that this appearance had nothing to do with the presence of copper. After most carefully washing the broken pieces of gold in nitric

acid, and after careful drying, I found that these portions weighed exactly *twenty-three grains*, which was precisely the amount which the coin-shaped piece of gold was proved to have lost!! The issue of this experiment upon chemically *pure* gold must satisfactorily prove that the disintegration of copper had nothing to do with the effects, as was objected in former experiments where copper was present."

CHAP. V.

PROSE AND POETRY.

1846—1855.

"To ——

"Broomfield, 185–.

"Dear Sir,

* * * * "I am well aware that many scientific men in London imagine that I am an individual having all my time to myself, and have nothing to do but to follow science without interruption, as leading the life of a perfect recluse. This is by no means the case, as in fact I have only a portion of the day to myself, at least which I can call my own. Sometimes I am deeply engaged for days together in unavoidable business anything but scientific. I have woods and plantations to thin out, waste lands to plant, farms to keep in order, and labourers to overlook; all this to do with a very moderate income, which requires the skill and address of a much more money-loving person than myself to manage without difficulty. Add to this my duties as a magistrate, a guardian of the poor, and consulted by all sorts on all sorts of business, obliged to come forward in county politics,— I really am a slave rather than a freeman.

* * * "It would take a volume to detail what I

have undergone, and I would not lead my unfortunate life over again for all the blessings which this world could shower upon me. * * * So that if I had made, or was to make the greatest discovery upon earth, I have had enough to counteract the pride of the proudest, and to feel that I am but like my frail fellow creatures — *dust and ashes.* * * * Many thanks for your kind communication from your Microscopical Society; it is highly interesting. From all such meetings I, as a country mouse, am wholly excluded, and have little scientific communication, save reading and writing, with any but my own thoughts.

* * * * * * * *

"Believe me, &c. &c.,
"ANDREW CROSSE."

"To ——

"London, August, 1848.

"Dear Sir,

"Your very kind letter addressed to me at Broomfield followed me to my address here, where I am staying to carry out my plans for purifying various liquids by electrical action, on which I have been on the whole pretty successful. I have been superintending the erection of apparatus for converting sea water into fresh. In order to effect this great change rapidly, it is necessary to distil the sea water *once.* It passes from the still sufficiently good for washing, but quite unfit for drinking; — a very inexpensive, simple, but effectual electrical apparatus being plunged into a cask of it, of any size, rendered it perfectly good, and palatable in one night. Water thus purified has kept in an open cask for 14 months, and is now as sweet as when first purified. This apparatus furnishes about 100 gallons in 24 hours. * * * The time is coming when the electric influence will work wonders, and cause an alteration far greater and more permanent than those political convulsions which are now agitating Europe.

"I thank God my health is in general good, and I am in good working order, at least for the *present.* I consider life as a dream, and mine has been a very painful one, with the exception of scientific pleasures, without which I know not what I should have done.

"Believe me

"Yours sincerely,

"ANDREW CROSSE."

"To John Sealy, Esq.,
"Bridgwater.

"21. George Street, Portman Square, London,
"July 18th.

"My dear Friend,

"Your affectionate letter reached me yesterday morning. * * * You paint very fervidly the beauties of our picturesque county, and seem to wonder at my long stay in this busy and broiling place. I perfectly agree with you as to the glorious magnificence of the scenery of the Quantock Hills; but where are those that rendered it dear to me gone? I feel that I have no home, that I am cast as a weak vessel on the billows of the ocean, and all spots of land are now much the same to me. When I am at Broomfield I have not a soul to speak to, servants excepted. * * * In fact I am quite desolate, and am now thrown on the aids of science, and the companionship of my friends; with the exception of your family, and one or two others, I have no intimacies in the country. * * * In this huge metropolis, there is a power of selecting one's acquaintances beyond that in any other situation, and moreover I have many very old and tried friends and schoolfellows here with whom I am on brotherly terms. They rejoice in my successes and grieve with my disappointments, and I fully reciprocate their feelings: nevertheless, I long to breathe, if but for a while, my mountain air, although every breeze of it is full of melancholy associations; and beyond this, I hope to lie in Broomfield churchyard, when it pleases my Maker to remove me from this troublesome world. * * * *

"Yours ever,

"ANDREW CROSSE."

The following are loose, unconnected scraps, taken from letters of Mr. Crosse's written at this period: they indicate the workings of his mind.

"I have no very high opinion of the infallibility of the human intellect, however and by whom exerted. Don't fancy, therefore, that I set any particular value on my experiments, otherwise than as feathers thrown up to see which way the wind blows.

"By the bye, I am of opinion that geologists and mineralogists often deceive themselves by ascribing to the action of fire appearances which are in reality occasioned by the *evolution of gas.*

"We must be blind indeed not to feel convinced of the predominance of DESIGN throughout the whole earth, in every ramification.

"I have every reason to believe that the electric action will be universally employed in vast variety of manufactures over the whole civilised world.

"It is better to follow nature blindfold than art with both eyes open.

"I have lived to see my favourite science standing high amongst the sciences of the world. Perhaps at some future period I may be permitted to see its boundless extension with an *unclouded* and unprejudiced vision.

"When misfortune oppresses, and the cares of life thicken around us, how delightful is it to retire into the recesses of one's own mind, and plan with a view to carrying out those scientific arrangements, with a humble hope of benefiting our country, improving our own understandings, and finding unspeakable consolation in the study of the boundless works of our Maker! Often have I, when in perfect solitude, sprung up in a burst of schoolboy delight at the instant of a successful termination of a tremblingly anticipated result. Not all the applause of the world could repay the real lover of science for the loss of such a moment as this."

On the subject of religious freedom he thus speaks: —

"I have ever been of opinion that the greatest insult which we can offer to our fellow creature is the impious endeavour to *force* him to accept as true the same religious creed as ourselves. The forcible intervention of one man with the religious belief of another is gross impiety towards our Maker, on the one hand, and the worst infraction of liberty of conscience on the other. The mere attempt so to act can only be productive of hypocrisy and of the stifling of all freedom of expression."

The following are some of Mr. Crosse's later poems: —

"SCIENCE.

"THE God who bade each element
Be formed, and it was done,
And raised the attractive power to link
The two or more in one;

"Who from the simples compounds made,
And bade the compounds roll,
Unnumbered globes, through Heaven's wide space
In one resplendent whole;

"Who willed that all around should change
That dwells in earthly frame,
Contrast to Him who was and is
Eternally the same;

"Who, set on this, our smaller sphere,
Man's fleeting race assigned
To lord it in a troubled dream
By virtue of his mind;

"Saw fit, through his Almighty ken,
Two opposites to blend,
The two extremes of good and ill,
The antagonist and friend!

"Half-blest, half-curst, this motley globe
Self-balanced whirls along,
And light and shade alternate play
Upon the right and wrong.

"And smooth and rough, and gay and sad,
And great and vile are found;
They mingle in our brightest hours,
In deepest night surround.

"Nor can the extent of human wit
The right from wrong divide,
For (such is the Creator's will)
They travel side by side.

"And as they soften into one,
The line where they unite
Yields not a stain its course to trace
Or point the wrong from right.

"So the pure stream glides from the land
Into the briny deep,
And fresh and salt together tossed,
In mutual eddies leap.

"But He who made the thorn so sharp,
Can heal the wound that's given,
And each new pang that rends the heart
May earn a brighter heaven.

"Oh! look around and bow the knee!
Not to earth's sceptred race,
The inflated froth of tyrant pomp,
The plaything of the base!

"Oh! look around and bow to Him
Whose works a language tell
Which vibrates to the inmost core
Of all who feel their spell!

"Nor does kind nature from the poor
Untaught her beauties shroud,
Nor deck her in her gorgeous dress
As tribute to the proud.

"For all alike the sun throws out
His unselecting beams,
For all the generous waters pour
Their multitude of streams.

"The tulip glows not on its stem,
To please the chosen few,
Nor do the birds for these alone
Their morning hymn renew.

"The terrors of the unsparing storm
Flash on the lightning's wing,
To smite, with equal fury hurled,
The peasant and the king.

"Yet though earth's thoughtless myriads
May Nature's wand obey,
May smile at flowers, or shrink at storms,
As charms or terrors sway;

"Can these, who scarce the surface skim,
Be struck with equal awe,
As in the mind of him who strives
To learn creation's law?

"On entering first the golden gates
Which Science opens wide,
The dazzling beams which issue forth
May wake his human pride.

"Forgetful of the crowds which throng,
Partakers of his toil,
He fancies he alone may snatch
The intellectual spoil,

"And proudly turns to those behind,
Who o'er his shoulders reach,
As though inhaling at a glance
What ages fail to teach;

"Like one that scorns the even path,
And bounding quits the plains,
To scale some rock whose sunny peak
In tranquil beauty reigns.

"A lesser point he soon surmounts,
On which, till then unseen,
Burst brake, and bog, and hideous chasm,
That sternly intervene;

"Whose silent language plainly tells
'T is fruitless to essay,
But with surpassing toil and art
To thread the devious way.

"Yet should he gain the hidden track,
And the bright summit climb,
Oh! what a flood of glory pours
Upon that point sublime!

"'T is so with Science; to be won
By toil must she be sought,
Nor can her smiles by gold or gems
(The hire of kings) be bought.

"Those who reap where they 've not sown
(A crafty race) she spurns,
And from the vacillating soul
With glance indignant turns.

"Impatient bursts at slow success
Suit ill, her bounds to pass,
Nor shall he reach the mental goal,
The slave of Fashion's glass.

"But him, the ardent votary
Who dauntless perseveres,
In spite of frowns from worldly friends,
In spite of adverse years,

"With not a breeze to waft him on,
With scarce a pulse to beat,
In concert with his hopes and fears,
Nor staff to prop his feet;

"Condemned to breast ignoble foes,
The canting bigot's art,
The tiger-spring of rival ire,
The head without a heart;

"The noisy war of vulgar throats,
The treacherous ambuscade,
The sneer of those who knowledge hate,
Insidiously conveyed;

"The cold, unsearching, wintry smile
Of fools, who measure man
As though the joys alone of life
Should fill his insect span,—

"Him will not Science laugh to scorn,
But, should he constant prove,
Expand his mind to high desires
And universal love.

"Expand, —and in a lofty road
By fanatics untrod,
Who, through mysterious ignorance,
Alone would worship God;

"Who close their sacrilegious eyes
On all that might be seen,
Save where they place their own dull lens
Them and their mark between.

"Such were the days when cautious monks
This globe their centre made,
Nor did Copernicus unscathed
The sacred ban invade;

"When Galileo dared not add
New worlds to those once known;
Heaven's empire might extend too far
For their terrestrial throne.

"E'en now, in these less rigid times,
Are breasts which none might tame,
Who fain would all dissentients chase
To feed eternal flame.

"But praised be He who with the will
Has not vouchsafed the power,
Or some self-glorifying judge
Whole nations might devour.

"Strange? that the boundless works of God
Such gratitude may win,
As that to trace their hidden laws
Should be denounced a sin!

"Yet, were such opposition swelled
To many a thousand-fold,
Should each smooth pebble on the shore
Start up a warrior bold,

"The hosts must bow their mail-clad head
When Science sweeps along,
And to the mighty queen of earth
Shout hail in choral song.

"Hail to the queen whose simple car
No trophied coursers lead,
Nor barbarous burst of brazen trump
Her conquering march precede!

"Hail to the queen who, though unarmed,
Invulnerable reigns!
But o'er no grovelling race of slaves
Her peaceful sway maintains!

"Who wears no chaplet round her brow,
With tear-bought gems entwined,
By mightier ornaments eclipsed,
The splendour of the mind!

"Who holds no titled baubles forth
To win rapacious hearts,
Nor terrifies untutored crowds
By State-invented arts;

"Who spoils not, to support false pride,
The many for the few,
Nor on the natural ills of life
Heaps up a thousand new;

"The comforter when woes assail,
The guide to loftier thought,
The interpreter of Heaven's fair works
For man's advantage wrought;

"Who bids us glance without — within,
And reverently kneel;
Without, God's empire view around,
Within, his temple feel;

"Who would, if all she might persuade
Direct their reason right,
And what is clothed in earthly clouds
Invest with sacred light.

"Before whose skill unwieldy strength
And ruffian force lie dead,
And tyrants fall ere yet they dream
That half their power is fled;

"At whose pure touch in dread dismay
Shall fierce Intolerance shriek,
And her unblessed abode afar
In kindred darkness seek;

"Whose voice the sleep of ages past
Shall burst from pole to pole,
And Freedom bid assert her rights,
And loose the captive soul.

"Hail to the queen, the mighty queen
Sent forth by Will Divine!
Whose rays, unlike the glare of courts,
Gain lustre as they shine!

"Alas! I sing as though my sense,
As far as it may reach,
Could pierce the mists that mock man's sight,
And laws unerring teach!

"As though the object which I see
Were to the thought conveyed
By some well known and certain road
On sure foundations laid;

"As though I could look inward on
The mystery of mind,
And from all doubt on every point
By reason be refined;

"Make all within, and all without,
In just accordance meet,
And each opposing principle
Cast down beneath my feet.—

"Thus might have dreamed some captive wretch
In solitary cell,
On whom the light of heaven alone
Through one close casement fell.

"If purple were the glass, the ray
A purple hue would shed,
And, if the medium gave the stain,
Would glow with ruby red:

"His eye the prospect spread afar,
Or red or purple sees,
Unknown the tints which deck the morn,
The verdure of the trees.

"Just so, — I grope my erring way,
Bound in life's fragile tomb,
And from the light which glimmers round
My thoughts their form assume;

"But *absolute*, unchanging truth
Escapes my warm desire,
Nor could my earthly shell sustain
Its concentrated fire.

"Perchance in some more favoured state,
From time and space set free,
My soul may rove, allowed to wake
To bright reality!

"To wonder at the shapeless mass
Of scientific pride,
And tremblingly adore the Power
Who draws the veil aside.

"Thus may it be, if such as I
Imperfectly might pray,
And may the unclouded blaze of Truth
Unite us in its ray!"

"ANSWER.

BY THE SPIRIT OF A DEPARTED POET.*

"YES! I was mad, thou say'st too true!
And so shall future ages tell,
When they that hapless fate review
Which bound me to an icicle!

"Yes! I was mad, to hope thy breast
One throb of pity could bestow.
Can Caucasus burst from his rest,
Or in his frozen bondage glow?

"Yet wert thou pure in thought and deed,
Wrapt in Religion's peaceful shrine,
But oh! though mercy was thy creed,
That boasted mercy was not thine!

"The wayward feelings of my heart
Thy stoic calmness could not share;
And since thou hadst not learned to smart,
Alas! thou knew'st not how to spare.

"Oh! hadst thou, stooping from thy pride,
Gently essayed some milder tone;
That soul which harshness turned aside,
Thy new-born kindness must have won!

* Byron.

"But my stern bosom none could tame,
No, not e'en thou of faultless grace!
Nought but affection's warmest flame,
Nought but undying love's embrace!

"Couldst thou not deem that they who feel
Intense the joys or woes of Heaven,
Whose hearts are not, like thine, of steel,
Have sometimes need to be forgiven?

"Can he, whom thrilling passions shake,
Secure the paths of virtue tread?
Grasp firm the right — the wrong forsake —
Alive to good — to evil dead?

"If such there be upon this earth,
He ne'er his mighty impress trod,
Nor from a mortal was his birth;
He must have moved a breathing God!

"Fate casts her sons in various moulds;
Here love or hate alternate reigns,
And there a sluggard current rolls
Its torpid influence through the veins.

"But 't was not thus by Nature made;
For thee my pulse was wont to beat,
For had a thousand tongues forbade,
I still had laid me at thy feet!

"How couldst thou seek, condensed in one,
A soul of fire — a tenor mild?
Expect the fervour of the sun
To glow without a spot defiled?

"Thou knew'st me ere my heart was thine,
I could not cringe to idle form;
The world's cold trammels were not mine,
But the wild waste and sweeping storm!

"And thought'st thou, when I owned thy sway,
To break or bow me to thy will?
Too sure thou didst mistake the way,
And in the bud affection chill!

"A saintly tread, or measured smile,
Perchance had touched thy chastened heart,
But ere I stooped to worldly guile,
'T was better, though 't was hard to part!

"Then triumph that thy firm design
Relentless sent me far to roam,
My country, friends, and hopes resign,
And sadly desolated home!

"Triumph that no tear might gush,
That marble firmness to disgrace,
Or soft emotion raise a blush
To mantle o'er thy yielding face.

"Yes! triumph that my muse is dumb,
And that though angry slanders rise,
No echo from the vault can come,
To chase their influence from the skies!"

TO AUGUSTUS.

TRANSLATION FROM HORACE.—BOOK 1. ODE 12.

"WHAT sage or hero first among his nation,
On lyre or pipe, O Clio, will you sing?
What God, whose deeds surpass all emulation,
Shall echo ring?

"Or in the glades of Helicon resounding,
Pindus, or Hæmus, with its tops of snow,
Whence the mad groves, to Orphean music bounding,
Fearlessly go.

"He, by maternal melody delaying
Streams in their course, and winds upon the plain,
Draws, by his skill, incomparably playing,
Oaks from their reign.

"What shall precede the monarch of the thunder,
Guide of the seasons, life-supporting Jove,
Who rules o'er all things an eternal wonder
Here and above?

"Nought is created his desert befitting,
Nought that exists can with his name compete,
Yet wisdom's Goddess, thron'd on high, is sitting
Close to his feet.

"Thee shall I sing, O Bacchus, bold in battle,
Thee, Virgin hostile to the beasts of prey!
Thee, Phœbus, fearful with the shafts that rattle,
Crowned with the bay!

"Pay to Alcides and the Twins devotion!
One lord of steeds, the other of the fight!
When they as stars recline upon the ocean
Trembling bright.

"Then falls the sea-foam from the rock retreating;
Hush'd are the winds, and fleet away the clouds,
And the fierce wave, that late the sky was meeting,
Its fury shrouds.

"Next shall I name Rome's founder never dying,
Numa, or Tarquin, proudest of the great!
Or Cato's soul, indignant and defying,
Rushing on fate!

"Regulus, Scaurus, Paulus high in story,
Heedless of life against the Punic foe;
Fabricius, these win the crown of glory
Verse may bestow!

"Him and Dentatus ever rudely faring,
With wise Camillus, all despising gold,
Stern hardship reared the noble mind preparing
In iron mould.

"As thrives the tree, unseen, yet never ceasing,
Just so Marcellus in his fame aspires;
Cæsar's bright star is as the moon increasing
'Mid lesser fires.

"Jove, lord of earth, who ne'er in sight abatest,
Cæsar (so fate bids) rests upon thy care;
Thou shalt reign first, and he of men the greatest
Thy glory share!

"Whither he quells the Parthian pretender,
Rightly subdued for ineffectual boast;
Indian, Seres, or remote offender,
Countless in host.

"Lower than thee, he rules the world's just centre,
Thou shak'st Olympus with thy weighty car,
Through profane groves thy lightning stroke shall enter,
Bursting afar!"

"RETRIBUTION.

"Oh thou, who fain would'st wake the slumbering storm
And earth's fair features and her race deform !
Wouldst cleave the mountain, and o'erwhelm the plain,
Roll sea on land, in land engulf the main !
Wouldst scourge the elements to mutual fight,
Dissolve all harmony, change day to night !
Wouldst wildly snap the secret links that bind
Each hostile atom in its place confin'd !
O'erturn the balance, scoff at Nature's laws,
And rear a barrier 'twixt effect and cause !
Wrench hope from man, and substitute despair ;
No voice to guide him, and no God to spare !
Spirit of Vengeance, in whatever world
Thy flaming banner was at first unfurled,
No despot thou, no demon uncontrolled,
Though fierce in carnage, — of unearthly mould !
Howe'er thou rage, whate'er thy fearful skill,
Thou must obey thy mighty Master's will.
Till He unloose, thy hellish power must sleep,
And while thou dreamest, man forget to weep !
But when the dread permission once is past,
Woe to the victim on thy fury cast !
Yet such th' unerring wisdom of thy Lord,
Not light the cause, that bares thy fatal sword.
Perchance to teach th' unpractis'd what is pain,
And smite himself, from smiting to abstain.
By wound on wound the grosser blood to free,
Till what is left flows in such purity

That heaven is half attained ere death destroys,
And pain 's the path to everlasting joys.
Or else, proud fiend, the high command is sped
To fill vast fields with hecatombs of dead,
T' infect the gale with pestilential breath,
And change its balmy sweets for dews of death,
Tyrannic kings and states from earth to sweep,
Who sow corruption, and corruption reap!
Then vengeance lights not his infernal flame,
But RETRIBUTION is thine awful name!
Great power who issuest from God's right hand,
When patience slackens, and when crimes demand,
To thee the kneeling slave extends his chain,
Though unprophetic tyrants plead in vain!
And thee the frantic mother's scream assails
When her lost child th' invader's spear impales!
And pillaged towns and bleeding hosts implore
Thy might to punish, though they hope no more!
Short-sighted man, whose grovelling thoughts scarce rise
Above the clod that underneath him lies,
Who, though he prates of heaven, no heaven can see,
Save in vile pomp and worldly dignity;
Who decks religion in the trash of state,
Sublimely mean, ridiculously great!
Who locks humility without his breast,
And rears his God in gorgeous fetters drest;
Around whose ken a sickening halo grows
Of gewgaws, fantasies, and splendid woes,

Of gold and purple, coronets and kings
(The worldling's altar, and the madman's wings).
Is it for these he sells his dearest prize,
And vaunts his bargain to deluded eyes?
For these he casts his reason to the wind,
And basely stoops to grosser food inclined,
Quits the bright path that leads to purer joys,
His country, kindred, and himself destroys?
What small beginnings end in mighty deeds!
For one man's vice how oft a nation bleeds!
A wish indulged soon gains resistless weight,
And what was first a thought o'erthrows a state;
But cursed is he who gives such thought the rein,
Who owns no impulse but his private gain,
Who, in the guise of patriot, prince, or priest,
Regards himself the first, — his country least;
Himself a wretched mercenary, *one*
By whose sole frown the million are undone,
A living libel on the wit of man,
But suffered since this suffering world began;
Perched on a vapory eminence, to be
The slave of slaves, the mockery of the free.
As the wide ocean yields in purest steam
Its watery bounties to the sun's fierce beam,
Which soon condensed descend in kindest showers,
And give to hills their groves, to vales their flowers;
Then in meandering rills their passage wind,
Next roll in stately rivers, when combined,
Till to the parent deep their floods they pour,
And what was borrowed without loss restore;

So does the good or ill which men bestow
To their own breast return a friend or foe!
A pitying glance from him whom realms obey
Can win a nation's heart, and pave the way;
So that which far transcends the lettered law,
Affection, duty, gratitude, and awe.
For one like this a thousand hearts would bleed,
A thousand, thousand swords would at his need
Around his sacred head form one bright sun
Eclipsing all that tyrants e'er had won.
But 'trust not princes' was the Psalmist's song;
Himself a prince marks the advice more strong.
The fools who trust or others would persuade
Make up some patterns to support the trade;
Some Alfred sprung in half-forgotten times,
When frauds were virtues, robberies no crimes.
'T was then they trusted, to their interest true
The saint bamboozled, and the monarch slew.
Yet grant some half score royal monsters bred
In sixty centuries, whose virtues led
Creation by the nose, and lured mankind
To fancy that the breed were more refined;
The odds is somewhat fearful; till of late
The people bowed, but did not calculate.
Would man look back upon his mind and see
Not what he is, but what he ought to be;
See God's bright impress stamped upon his soul,
Zeal to excite and reason to control;
Conscience to check when blandishments entice,
Or burst in thunders at committed vice;

See beyond these, beyond this wintry dream,
The golden heights of life eternal gleam ;
Say, would he sacrifice the blessed chance
For all that might his earthly hopes advance,
To cull the fruits which ripen for the proud,
An hour of thankless homage, and a shroud ?
Ye mighty bards, who lent your thrilling strain
To vaunt as God-like deeds that were profane,
Who pinnacled on throne of empty fame
The wretch whose crimes should sink obscured in shame!
Who, in the clash of arms and rolling car,
The neigh of steeds and panoply of war,
Gave reins to fancy, judgment overturned,
Ambition deified, and justice spurned ;
Saw not that glory, treacherously bright,
Shed as the icicle delusive light,
But poured the magic numbers of your song
To lull admiring ages to their wrong !
How could ye stoop to herd with vulgar slaves,
And tune immortal verse for thankless knaves ?
Were there not they whose wit might better grace
The bloodstained trophies of a plundering race,
Might raise a discord for discordant deeds,
And for the laurel weave a wreath of weeds ?
But must ye, whom to name in ought but praise
Were almost sacrilege, attune your lays
To an ignoble theme, and usher in
With such stupendous pomp such damning sin ?
Think not, ye spirits of the mighty few,
Your slave presumes to drag your faults to view ;

Or that in madness he could hope to soar
Within the blaze of those whom worlds adore!
Would conquering heroes their own praises swell,
Few to posterity their feats would tell.
So clear is justice to the weak or strong,
The very child points out the right from wrong,
Though reasoning senates close or ope their eyes
As right or wrong may gain the better prize.
Mark the sly tempter, how he plays his part,
And strains his viewless net to catch the heart.
In untaught climes a ruffian method tries,
And to be cruel, feigns, is to be wise;
The burning pile, the tomahawk, the shaft,
And thirst of vengeance there compose his craft.
Where education sways he is more nice,
With balmy sin and soft voluptuous vice.
Not all at once he swoops upon his prey,
But steals the senses to ensure the way;
Holds forth some glittering bauble to ensnare,
A fortune, mitre, crown, or yielding fair,
Draws the charmed eyes to view the sunny side,
And what is dark contrives in mists to hide;
Wraps all the senses in poetic glow,
And spreads a flowery veil o'er human woe;
Schemes in deep learning, and in science lurks,
In grave religion tenfold mischief works;
Then as his plans mature he grows more bold,
And the snare, once too slight its prey to hold,
Now drags the wretch in adamantine fold.

Each virtue outraged and compunction flown,
Dire Retribution claims him for his own ;
Claims him e'en here (too just to wait, oppressed
With pains prepared to pierce another's breast."

"MAN.

" Oh, new-born man, thing of a day,
Poor helpless mass of living clay,
Who moan'st thine infant hours away,
What being would envy thee?

" Slave to the scourge and petty vice,
To pedagogue and prejudice,
When chains of boyhood weave new ties,
What being would envy thee?

" Scorched by the blaze of mad desire,
When the veins glow with youthful fire,
And Folly draws her victim nigher,
What being would envy thee?

" By courtiers, lawyers torn asunder,
By all who deem thee worth the plunder,
Taught at false friends to grieve or wonder,
What being would envy thee?

" Tortur'd by vile ambition's sting
The sycophant's applause to win,
All puff without, all void within,
What being would envy thee?

" Ground to the earth by sordid lust
Thy soul low grovelling in the dust,
Thy God abjur'd, in dross thy trust,
What being would envy thee?

" And when thy strength has pass'd away
While reason faints into decay,
And death now claims his worthless prey,
What being would envy thee?"

"THE WORLD AND I ARE TWO.

" When waking from the trance of life
Its changes I review,
I feel and sadden at the thought,
The world and I are two.

" I might have loved, 't was glitter all,
Light, wavering, and untrue;
My love return'd unto my breast,
The world and I were two.

"I would have shared my inmost soul,
But kept my thoughts from view,
For soon I found it, to my cost,
The world and I were two.

"I would have grasp'd a friendly hand;
Ah me! what could I do?
So cold, so heartless, and unkind!
The world and I were two.

"When love and friendship thus were fled,
Respect might still be due,
But each was swallowed up in self,
The world and I were two.

"And we are two, and so shall be
Till life's unwound her clue,
And then without a pang I'll cry,
The world and I are two."

"TIME.

"Time rolls on, till all are swept away,
The wise, the blind, the preyer, and the prey!
Still there are those who, born surpassing great,
Refined by thought, or purified by fate,

Stars of their age, and lights amid a storm,
Framed to excel as much as some deform ;
Nature's choice flowers, whose fragrant mem'ries bloom
When long their stems have withered in the tomb ;
Still there are those who grace the historic page —
Bard, hero, patriot, senator, and sage —
On whom, exalted to a height sublime,
Death claims no power, no power has ruthless time ;—
Such are the names my much-loved country sees
Entwined amid her glorious destinies !
Such are the suns whose beams shall matchless shine
When nought of Britain lives but what's divine!"

About this period, 1849, Mr. Crosse spent most of his time in London. It was in town that I made his acquaintance. I remember well the first time I saw him. It was at a dinner-party at the house of a mutual friend.

When young I had always been intensely interested in Mr. Crosse's experiments in electrical science; I had cut out scraps from the newspapers that made mention of his discoveries, so that it was with no common feelings that I looked upon the man whose power in wielding that mysterious agent, electricity, had so excited my imagination. I sat opposite to him at table, and listened, as he talked lightly, pleasantly, upon the subjects of the day. Perhaps he was hungry — I was disappointed — he did not speak of electricity.

Later in the evening I found myself one of a group of listeners near Mr. Crosse. Something had awakened an old memory of fancies, which he had aforetime linked into verse. He was reciting some lines of his own poetry; but it was not recitation, it was rather the emphatic utterance of a torrent of thoughts, that fell in rhythm. How his words stirred one's heart, as he changed from the deepest pathos to a burst of earnest patriotism; or, in language simple as the wild flowers, whose scents and hues he summoned round him, he led you in fancy to those wild hills, *his* home, and *yours*, for the time, for while you listened you felt what he felt. I had expected to find what I reverenced — a follower of science: I found what I worshipped — a poet. Nor was I alone in my enthusiasm. I have rarely or ever seen a person who made so deep an impression at once, even on those whom he casually met in society. There was an entire earnestness and reality in his character, which was appreciated even by the most artificial. The enthusiasm with which he threw himself into subjects connected with his favourite science infected all listeners, who were almost invariably carried away by the unstudied eloquence of his descriptions of Nature's wonderful laws, and his sanguine anticipations of the successful imitations of art.

When I saw Andrew Crosse again it was in his own house. And here he was in the threefold character of host, country gentleman, and philosophic experimenter. He used to say, "I think hospitality so essential, that I don't call it a virtue, but I do call the absence of it a great vice;" and, "If my greatest enemy was to come to my house, I would ask him to dinner, and call him out afterwards." The first I am sure he would have done, the last I am sure he would not; for under the most galling provocations, he *never injured another.* He spoke strongly, often too strongly. He was sometimes mistaken by those who could not understand that an impetuous nature like his had not that *caution* that makes some people seem, but only *seem,* more mild. He could forgive freely, and, what is more, bear to be forgiven. I might quote his own lines descriptive of a poet's character: —

"Couldst thou not deem that they who feel
  Intense the joys or woes of heaven,
Whose hearts are not, like thine, of steel,
  Have sometimes *need* to be forgiven?"

* * * * * *

"How couldst thou seek condensed in one
A soul of fire — a tenor mild?
Expect the fervour of the sun
To glow without a spoil defiled?"

At Fyne Court Mr. Crosse was surrounded by a perfect chaos of apparatus. Certainly the old house had more a philosophic than a domestic air about it. The family plate was occasionally called on to make contributions to the crucible, which, with the aid of the laboratory furnace, converted tea-pots, tankards, and old-fashioned spoons into *chemically pure* silver in a very short space of time. A great deal of the glass and china of the house was not suffered to remain in vulgar use, but was dedicated to nobler purposes, and was formed into batteries or other electrical arrangements. The rooms generally seemed in a process of resolving themselves into laboratories or other kinds of scientific dens. You were perfectly comfortable, perfectly at home, under that hospitable roof; but, to speak in geological language, the house appeared to be rather in a *transition* state. Lady Lovelace, who was often a guest at Fyne Court, used to say "that the dinner hour was an accident in the day's arrangements."

There is a story that after Mr. Crosse had finished building the sixth or seventh furnace in his laboratory, he said, with an air of great satisfaction, "I consider *now* that my house is thoroughly furnished."

The old place had got into a terrible habit of wanting repair. In a storm, the slates of the roof seemed as if positively electrified, and flew right and left in mutual repulsion. When I first saw Fyne Court it was besieged by an army of masons and carpenters; Mr. Crosse's large philosophical room had fallen down. Dr. Buckland, at the inauguration meeting of the Archæological Society at Taunton, gave a humorous description of this accident, attributing the circumstance to the effects of misguided lightning, which the electrician was supposed to have trifled with. The facts of the case, however, were more prosaic. A bad architect and a dishonest builder were the real foundation of the mischief. In the autumn of 1849 this room had just been rebuilt; but in the midst of all these disturbances numbers of batteries were at work in different corners of the house. You were taken, perhaps, to an underground cellar to see the progress of an agate that was forming in an electrified solution. Shortly afterwards you might find yourself in a mysterious chamber, "dark as Erebus, black as night," excepting where it was dimly lit by the magician-looking lamp, carried by the philosopher himself. There was no sound in that silent room, except the ceaseless and regular dropping of water, an electrical arrangement for favouring the growth of crystals, in imitation of Nature's processes in her subterranean caverns. Unlearned as I was, I was much struck with what appeared to me the inexplicable mass of materials and apparatus; their shape, form, qualities, names, and uses were alike a puzzle to me; they could not remain so long, for Mr. Crosse's simple and ready explanation would have made their purpose clear to the most common understanding. One hardly knew at times whether to be most amused by the simplicity and ingenuity of his adaptations, or to admire reverently the wonderful results which proved these arrangements to be closely analogous to Nature's own workings.

Mr. Crosse's conversation was full of anecdote, and very amusing. He recounted the visits he had received from various celebrated people, and their sayings and doings.

Professor Sedgwick, I recollect, with some considerable amount of poetical licence, had described the arrangements for conducting atmospheric electricity at Broomfield, as bringing into the philosophical room streams of lightning as large as the mast of a ship. The history of Dr. Whewell's visit to

Broomfield was full of incidents, of which he gave me a humorous description some years after.

On one occasion Baron Liebig, Drs. Buckland, Daubeny, and Lyon Playfair, together with some other men of science, paid Mr. Crosse a visit. From letters written afterwards, they all appear to have been much pleased with their day at Broomfield. An amusing circumstance occurred in connection with their visit. As this party of strangers passed through the neighbouring town of Bridgwater their appearance excited great surprise, indeed some consternation, more especially when it was rumoured that they were about to pay a visit to Andrew Crosse, whose science, poetry, and politics were alike incomprehensible to the rustics round, and had gained for the poor electrician something more than a doubtful reputation in the neighbourhood, "The prophet's own land." His learned friends fell under the shadow of his fame; and, after many speculations, the police came to the conclusion that they were dangerous people, Chartists every one, and that no good would come of their visit. Dr. Buckland was particularly amused when he heard the story afterwards.

Baron Liebig's inquiries were harmless enough; he was interested in knowing the manner of making Cheddar cheese.

While at Fyne Court, in looking over Mr. Crosse's collection of factitious crystals and other electrical results, he remarked immediately the *sub-sulphate* of copper, exclaiming, "Ah, here I see you have formed a *new mineral.*"

I remember Mr. Crosse telling me of a morning that Mr. Sydney Smith and Dr. Holland spent with him at Broomfield. He performed an experiment which neither of them had seen before, namely, the formation in the electric circuit of ammoniacal amalgam, as discovered by Sir Humphry Davy, who found that mercury could be made to swell to twenty or thirty times its original size in a weak solution of carbonate of ammonia.

By a slight change in the arrangement of the same experiment, Mr. Crosse caused mercury, in connection with the negative pole of a voltaic battery, to float on water. The instant the electric circuit is broken the mercury returns to its original size, and falls immediately to the bottom of the vessel. This was a favourite experiment with Mr. Crosse when he gave lectures, which he frequently did for the benefit of mechanics' institutes and scientific societies in the neighbourhood.

Mr. Crosse's recollections of Mr. Sydney Smith included many anecdotes of that gentleman's witticisms. I remember his telling me of a charity sermon commenced by the Dean of St. Paul's in the following manner: "Benevolence is a sentiment common to human nature. *A.* never sees *B.* in distress without wishing *C.* to relieve him."

On one occasion Andrew Crosse was sitting opposite Sydney Smith at dinner; the latter was jocosely remarking that he would have bishops in every corner of Great Britain, in every island round about her shores. "I would have," said he, "a Bishop of the Flat Holmes, and a Bishop of the Steep Holmes." *

"That would be a great advantage, Mr. Sydney Smith, for they would be surrounded by their sees" (seas), observed Mr. Crosse.

"I will place you next * * *, one of the most agreeable talkers in London," said his hostess one day to Mr. Crosse, when he was dining with * * * at his residence in Surrey. "At dinner I made the following observation to my neighbour," said he, "which seemed to touch the chord of memory; my observation was, 'I think there is more romance in real life than was ever imagined by the most extravagant writer of fiction.' * * * fully agreed with me, and then poured forth a host

* Two rocks in the Bristol Channel, boasting some half-dozen houses.

of anecdotes of his own personal experience of romance in real life." A lively interchange of stories then took place: Mr. Crosse told me again of several of the anecdotes which he himself narrated. I am grieved that I cannot remember them distinctly enough to be their chronicler, but I recollect well the changes from grave to gay, pathos which almost drew tears from my eyes, and startling humorous absurdities that sent me into fits of "inextinguishable laughter." But it is a vain effort that of endeavouring to give even an idea of any man's conversational powers. The time, place, and antecedents can never be reproduced. It would be a boon to literature if table-talk could be photographed, more especially if the plates could be made cunningly sensitive to wit and humour *only*.

Mr. Crosse's serious conversation was peculiarly suggestive. It was impossible to listen to his very original views upon all subjects, without feeling a host of fresh thoughts burst upon the mind; he lifted you up mentally, as a tall man might a child, and you saw clearly a more distant horizon. His memory was very remarkable. He never forgot any of his own experiments, and this was partly the reason why he was so indifferent about noting them down, and hence much valuable knowledge has been lost to the world. He would detail the minutest particulars about electrical arrangements made years before. He could give the history of every crystal he had formed; nor were the phenomena of his atmospheric exploring wires forgotten, though they had recorded the alternations of so many summers and winters. His memory was equally good on any subjects that interested him. It is no exaggeration to say, that of his own poetry he could recollect without hesitation some thousands of lines; he would also quote extensively from the poets of antiquity, and also those of modern times, more especially Milton and Pope. This delightful faculty rendered him a charming companion at all times, even to those who knew nothing of science. He had no idea of display: I doubt whether he ever understood even the small vanity that some people struggle so ineffectually to hide. He spoke because he was full of his subject; but it was to elucidate the facts, not to represent himself or his own merit in the development of a truth. His individuality was remarkable: you could never forget Andrew Crosse; —you could never confound his expressions; his opinions, right or wrong, were his own. Never was a man less of an egotist; yet, if I may be allowed the paradox, he had the egotism of genius, which involuntarily makes all things its own and pay tribute to its sovereignty. Every thought that passed through the alembic of his brain was stamped with his own image, it could not be otherwise; he saw things on the speculum of his *single* mind.

In poetry, Mr. Crosse found expression for the intensest feelings of joy and sorrow. The *latter* is the muse's nurse, as Shelley says —

> "Men are cradled into poetry by wrong,
> And learn in suffering what they teach in song."

Mr. Crosse did not affect much refinement in his poetry; he said in a letter to a friend, "The age is too refined already." His compositions were written rapidly, and rarely corrected; except in downright fun, he never wrote what he did not feel. His poetry is full of generous and noble sentiments, of pure and delicate feelings; but criticism does not belong to *me*.

"To * * *.

"Broomfield, Dec. 4th, 1849.

"My dear * * *,

"Your most welcome communication reached me yesterday. * * * * * * Did you ever observe in an early summer sunrise, just before that glorious orb made his appearance, a stratum of sky above the horizon of a pellucid, unearthly golden *green* hue? It is rare, but sometimes to be seen. I feel surounded by a similar kind of halo — a mixture of the *earthly* with the *unearthly;* such is my waking dream: — all sorts of real and unreal things pass before me, and I feel as though living in the days and country of the Caliph Haroun al Raschid. I fear I have far too much romance in my unfortunate composition for the stern realities of these frigid times. This I cannot help, as I was cast in such a mould. * * * No tongue can describe the acuteness of feelings with which I *lately* read some beautiful lines in the title page of an old prayer book, written by my mother, and given to my late sister. My dearest mother was a pure, simple, and innocent being, with a lofty and generous mind. * * * * Deep feelings are awakened in my soul, which no language can enable me to pour forth, to which language is a *barrier.* I have often thought how infinitely *music* surpasses poetry in expression. The last, it is true, in its loftiest flights is sublime; *but* it is definite, hedged round with a fence of words, often unmeaning, more often inadequate to the purpose, and never, never capable of allowing the *thought* to soar to its utmost ambition; whereas in music the transition from the tender to the sublime, from the low to the high, is easy and gradual. Its powers are stupendous. It is sufficiently measured and guarded by time, and yet it is essentially indefinite and free to mount to the loftiest altitudes. Alas! I look back upon my own mind, so full of all imperfections. I would be what I am not: I would grasp the good, and flee from the evil; but *here* they are inseparable. I would express to you what I feel, but I am struck down to the ground; a clod! * * I am called away by worldly but necessary matters, and I must wind up this tiresome scrawl with praying my great Maker to bless you and yours, here and hereafter, and remain

"Your ever sincere friend,
"ANDREW CROSSE."

"15. Charles Street, Manchester Square, London,
"December 16th, 1849.

"My dear * * * *,

"Your kind communication reached me three days since. I will state my proposed movements as far as I can at present. In the first place I have engaged, on my return to the West, to pay a visit to your father; from thence I shall go to Winslade, to stay a short time with my good cousin Henry Porter. I have also partly engaged to stay with some friends in Somersetshire, and I must visit my old home at Broomfield, to mark a great quantity of young trees there, as they are occasioning great injury to the more valuable trees which they oppress with their branches. After this roundabout I must fly back here, to attend to the scientific experiments which I am carrying on in London. * * * You ask me about my health: on the whole I am moderately well, but have suffered much since I have been here; the drizzling fogs, continued rain, with steaming heat and a host of disagreeables have much annoyed me. When, however, I look around, and find myself an isolated animal,—my best and dearest friends gone, myself standing all but alone in the gap, ready to be carried off by the next discharge of our common foe,—I feel that I have little to do in this world, but to brace my mind to submit *humbly* to the fate which my wise Maker may ordain for me. With respect to your opinion on *music*, I agree to a great extent; but I had not time in my letter to you to say all I intended. There are two sorts of music and two sorts of performers. The one, light, frippery, worthless balderdash, putting one in mind of Pope's witty remark —

"'Light quirks of music broken and uneven
Make the soul dance upon a jig to heaven.'

"Such are the modern performances of a boarding school Miss in her teens, or even the highest modern fingering in the most fashionable drawing rooms; noise without effect, execution without taste, *worthless* and *soulless.* The other, *grand*, awful, magnificent, as swelling out in Handel's godlike choruses, or dying away in the soft sweet strains of the inimitable Corelli. But where are the performers? *Dead, gone;* fled with the last trace of that sublimity which they have carried with them. Your trashy performers are men and women of no soul, sickly and unmeaning sentimentalists; but the few, the very few, who look back on the dream of past sublimity in this wonderful art with delight, are in general a widely different class, whom I will not praise here, as I myself am one of them. I may, however, say this of my late dearest and most honoured brother, that he who was the best performer I ever heard, who made one's inmost soul thrill by his powers, was the greatest lover of *truth* I ever met with. Please to observe that it is far from my intention to find fault with those who have no ear for music. My father was one of these, and his soul was the very seat of honour. Besides this, I am compelled, and with grief, to allow that Handel himself was a selfish glutton; we therefore must take man as he is, a compound of angel and devil. You say that *my* character is 'very peculiar.' I am aware of it; though, like my own species, I know myself *but little.* Still I know or believe that I am composed of two extremes in most things. I do feel most strongly, too strongly sometimes for my own comfort; but this is somewhat balanced by a power of reflection (I speak humbly) that enables me after a time, I hope, by God's aid, to correct and control any exuberance of wild thoughts which might otherwise carry me away. I believe that I am capable of *friendship*, and perhaps to a somewhat unusual extent; but to be called forth it must be *reciprocal.* Your remarks concerning me, my dear * * * *, are far more favourable than I deserve. Indeed, where I have been traduced and abused by those who do not know me, I console myself with the reflection, that I have been too often, far too often praised, and I let the excess of the *one* contend with the exuberance of the other. Alas! I have nothing to boast of but misfortune. I write to Robert by the same post, giving him the same account of my movements which I have given you. You observe that you were glad I was not alone at Broomfield during my late sojourn there. I never am alone when I have my faithful *dog* Greg with me. Where shall I find a heart one hundredth part so true to me as hers?

What a dreadfully long and tiresome scrawl have I been boring you with! I am always glad to hear from you when you have time to write, and at your perfect convenience, but don't trouble yourself unnecessarily, as I hate giving one atom of trouble, more than any one can imagine, to any human being. The day of Queen Adelaide's funeral was a very gloomy one here—the shops shut, bells tolling in all the churches, including the deep roar of St. Paul's, heavy guns thundering through the streets, and all the et-ceteras of royal pomp, telling the tale of the death of one poor exalted creature, while the destruction of tens of thousands who pass unnoticed from the earth, *one in every second,* is not dreamed of. Such is the world! a dream of vanities, fripperies, and nonsense! The late queen was, however, a *very good,* but not a *great minded* woman. I am not about to find fault with her, nor any other of my unfortunate species. God have mercy upon us all!"

"Yours very sincerely,
"ANDREW CROSSE."

"15. Charles Street, Manchester Square,
"London, Jan. 8th, 1850.

"My dear * * * *,

* * * "I am suffering much from nervous attacks, which I have had at times, more or less, ever since the age of fifteen. When about the age of thirty, I consulted Sir Anthony Carlisle; his answer was, 'It is mostly indigestion; eat dry biscuits, and drink water alone.' I did

so, and became vastly better; but at intervals they come back. The most agonising toothache (from which I have suffered much) is a very great pleasure when compared to them. What I have endured from them is only inferior to what I have suffered from the death of my nearest and dearest relatives; and then, during those overwhelming afflictions, I have given myself up entirely, as a useless incumbrance upon earth, and bowed to my Maker, mentally speaking, in dust and ashes. I hope, however, that something good may be gleaned from all this, as I am taught by my own feelings to pity *all* my fellow creatures, and to wish them well. I was going to add free from pain; but no—not quite free; with only enough to refine them from the grossness of humanity, and fit them for a purer and happier condition. God grant that this may be my case! It is very strange that, with the exception of this terrible feeling of nervous attacks, I hardly ever was better or stronger in my life: but 'what can't be cured, must be endured.' This is my mental motto. Forgive me for scrawling so much about *myself*, but I write as I feel. I agree with you entirely in your remarks upon solitude: it is then, and then only, that one can commune with one's own mind: 'Look inward,' as that glorious Milton says. I do not regret the years upon years that I spent at my dear Broomfield with my family—almost alone, save with them. * * * For those who are gone I would have suffered martyrdom; to rejoin them is my daily prayer. My good and wise brother used to say, 'that we should mix sufficiently with our fellow creatures to improve our minds, and be sufficiently in retirement to reflect on what we have heard from them.' I likewise agree with your opinions on the state of those wretched prisons called 'Union Houses,' which should be more aptly termed Disunion ones. Every kind of education, from the throne to the cottage, is morally, intellectually, and physically wrong. There is so much evil upon earth, that I am almost led sometimes to agree with a kind old friend of mine here, who believes that *this* is a place of temporary punishment for sins committed in a former state. After racking one's brains to endeavour to account for that which surrounds us, I think we can come to no other conclusion, than that expressed by my brother the day before his death: 'We know nothing of the past, the present, or the future; we can only pray.' I wish you had been acquainted with him; his mind was one of the most powerful that I ever met with; and he had the purest love for his fellow creatures. What a loss his death has occasioned to me! * * * * * *

"Yours very sincerely,
"ANDREW CROSSE."

"15. Charles Street, Manchester Square, London,
"January 12th, 1850.

"My dear * * * *,

* * * * * * * * *

"The streets are a sheet of slippery ice, horses falling in all directions. The night before last I was at a ball given by Fonblanque, the Editor of the Examiner. 200 persons were there. I was at another party the same evening, filled with wild and talented people. Yesterday

I dined at a party where I was invited to meet a young Hungarian nobleman, just imported, and a strange looking German. These people were all of the *haut ton*, and the Hungarian, who is very rich, talked of buying a stud of English horses, &c., but not one word of his poor enslaved country. I could not stand this, and I proclaimed myself a warm admirer of intellectual republics. This brought on a discussion, in which we all ended, as usual, by each retaining his own views. We all, however, agreed in one thing — a thorough detestation of *priestly bigotry*. All the intellectual agree in this one point. After leaving this party, I proceeded to another, where, there being an assortment of young animals of both sexes, there was, of course, a dance — or rather a variety of those eccentric motions. In the midst of these fancies, I slipped out, and got to bed before midnight, and this morming I feel vastly better. It is strange that I feel perfectly at home, whether in the midst of a compressed party of utter strangers in a London drawing room, or on the tops of my dear Quantock Hills; not so in country towns. * * * *

* * * * * *

You often talk of me as a philosopher. In the Greek sense of the term — a lover of wisdom — I am so, but a very humble and imperfect one, knowing well that little is to be gleaned here, but praying devoutly that I may at some time be permitted to snatch a glance at what *true knowledge* is. My soul would roam from sun to sun, from planet to planet, — inhaling every successive instant fresh portions of the Omniscient. By the bye, it is said that there is a spot at present on the sun's disc, of very unusual size. I am sure no Londoner could have made this discovery, as the sun has been long veiled from this place, under a canopy of murky fog. I met the other day at a dinner party the two Bretts — the electro-telegraph men, who are now preparing a sub-marine telegraph to unite Dover and Calais. As to poetry, I must not dream of it yet. * * * Although * * * and I are east and west in religion and politics, we are closely united in friendship. This is as it should be. * * *

"Yours very sincerely,

"ANDREW CROSSE."

I saw Mr. Crosse again in the early part of 1850. Amidst the recollections that crowd upon me, I remember particularly his conversation on one occasion. He was staying at the house of some relatives of my own, a small party of gentlemen had been asked to meet him. Some of them were men of science; all were interested in the subject.

One of the party had been speaking to him on his favourite science, and the possible effect of the electric agency in the *original* distribution of matter. He was standing near a table, he had a roll of paper, I recollect in his hand, with which he had been exemplifying some cylindrical arrangement for voltaic batteries; but these details of man's puerile imitations of nature, were at once passed over when the importance of the subject rose to a consideration of the electric influence as directed by the Creator himself in his primal work.

Mr. Crosse's whole face and figure seemed lit up with enthusiasm; his very tone of voice and words rose to eloquence, as he described the philosophic conception of chaos. He dwelt on the idea of the two simple elements uniting together electrically, and by their changing combinations forming the endless variety of substances which make this world a scene of beauty, a place habitable for man, supplying resources for the support of succeeding races, writing on tablets of *Time strata* the wondrous play of atoms, obedient, ever obedient to the law of CHANGE. "Whether we contemplate," he said, "miles of rock teeming with organic remains, or regard the changes in the vegetable kingdom, still are we filled with wonder and with awe. Mighty forests are hurled upon their native soil; centuries pass away, they are covered with mould, and blacken in their tombs; thousands of years succeed, and the buried trees are mineralised, and become vast strata of coal; the coal is consumed to create steam, and that which was once a seed lends its aid in impelling mighty ships across the ocean. Its carbon is given to the atmosphere, and again becomes a component part of new forests. THIS IS THE LANGUAGE OF THE DEITY! That Great One has ordained that nothing should be lost; that even principles apparently the most adverse should work His will, should contribute their quota to the demands of nature, should afford instances of such wonderful adaptation to the wants of man that all the collective wisdom of the universe could never have imagined. Great are the changes around: trees are converted into silica and chalcedony, probably by *electric transfer;* fissures are formed in the earth, their sides lined with minerals, we believe by the *negative* influence of electricity; the tender blade of grass, the gigantic oak, probably thrive by the electric power. Light, heat, and magnetism resolve themselves into its nature; the gases are held together and separated by its power. Its affinities permeate all matter. The aurora palpitates in obedience to its laws; its voice is echoed in the thunder-cloud; its presence is seen in the lightning. Metaphorically speaking, electricity may be called the right hand of the Almighty."

Alas! This gives no just idea of that spontaneous burst of eloquence: the grandeur and vastness of the subject found expression more in his looks and gestures, even, than in his words. Mr. Crosse paused a moment, and looking round, he was apparently, for the first time, conscious that the whole party were listening with rapt attention to every word that fell from his lips. He looked disconcerted, and with characteristic shyness and humility, turned to his nearest neighbour, and in a subdued tone continued the details of the voltaic battery, the subject in which he had interrupted himself.

Some time after this I heard Mr. Crosse give a public lecture on "Atmospheric Electricity," at the request of the members of the "Scientific and Literary Institutions," at Exeter.

On these occasions of public speaking he commenced generally with some amount of hesitation, nor would this wear off till he forgot himself in his subject. The fault then was, that so many thoughts crowded upon him, that he could scarcely give his audience time to follow his explanations of one phenomenon before their attention was demanded for another, and sometimes irrelevant, branch of the science. He was always particularly fortunate in his experimental illustrations. I hardly ever knew them fail. This was owing to the delicacy and precision of his arrangements. "He manipulated like a cat," said Mr. Kenyon.

Among the many complimentary notices that appeared of Mr. Crosse in the newspapers of the day,

I was about to select some verses which had always pleased me, especially one line which says—

"I see thee stand a poet in thy acts;"

but on consideration they seem to me too eulogistic to be consonant with the character of him whose brief memoir I am writing. I will, however, subjoin some lines addressed to him by his friend Walter Savage Landor.

"TO ANDREW CROSSE.*

"Although with earth and heaven you deal
As equal, and without appeal,
And bring beneath your ancient roof
Records of all they do, and proof,
No right have you, sequester'd Crosse,
To make the Muses weep your loss.
A poet were you long before
Gems from the struggling air you tore,
And bade the far-off flashes play
About your woods, and light your way.
With languor and disease opprest,
And years that crush the tuneful breast,
Southey, the pure of soul, is mute!
Hoarse whistles Wordsworth's watery flute,
Which mourn'd, with loud indignant strains,
The famisht Black in Corsic chains:
Nor longer do the girls for Moore
Jilt Horace as they did before;

* Written some ten (I think) years since.

He sits contented to have won
The rose-wreath from Anacreon,
And bears to see the orbs grow dim
That shone with blandest light on him.
Others there are whose future day
No slender glories shall display;
But you would think me worse than tame
To find me stringing name on name:
And I would rather call aloud
On Andrew Crosse, than stem the crowd.
Now chiefly female voices rise
(And sweet are they) to cheer our skies,
Suppose you warm these chilly days
With samples from your fervid lays.
Come! courage, man! and don't pretend
That every verse cuts off a friend,
And that in simple truth you fain
Would rather not give poets pain.
The lame excuse will never do,
Philosophers can envy too."

Mr. Landor frequently urged his friend to publish his poetry, and sometimes Mr. Crosse half promised he would; but his verses always remained on scattered scraps of paper, till I collected together the most finished of his poems, and transcribed them with some kind of order.

Mr. Crosse spent the spring and early summer of 1850 in London, where he entered much into the scientific and literary society of the metropolis.

On the 22nd of July, in the same year, we were married in London, at the Church of St. Marylebone. * * * *

After our marriage, Mr. Crosse resided more in Somersetshire and less in London. The old house at Broomfield had been long neglected, and many home duties claimed his attention.

Before we settled down for the winter, we spent some pleasant weeks in the north of England, and when at Ambleside received much kind hospitality from Dr. Davy, the brother of Sir Humphry Davy. We met, at his house, a brother and niece of Maria Edgeworth: in the latter, Zoë King, my husband found an old West-country friend. From the amiable family of the Davys, we heard many anecdotes of those celebrities, living and dead, who had made that immediate neighbourhood classic ground.

The names of Scott, Shelley, Wilson, Coleridge, De Quincy, Harriet Martineau, had "a local habitation" there. We saw the houses of Wordsworth and Southey: we stayed some days at the Swan Inn, at Grasmere,—that little wayside hostelry that Wordsworth had honoured with a mention. Every turn of the road, every fresh grouping of mountain and lake, recalled some line of those who had so sweetly sung the beauties of the "Poets' Corner" of old England.

My husband was delighted with the scenery in the north; but I think he enjoyed beyond everything else the storms and the minerals. He always went out armed with a hammer, and the numerous quarries and mines gave ample scope for its use. We generally came back from our walks loaded with stony treasures; and on our way, if a thunder-cloud loured on the head of Skiddaw, he would watch the dark rolling masses with mingled feelings, suggested both by poetry and science. He would interrupt himself, when detailing the electrical reasons for the concentric zones in a thunder-cloud, by bursting forth with Byron's description of a similar scene in another land,—

> "—And Jura answers from her misty shroud
> Back to the joyous Alps that call to her aloud."

While at Ambleside, we were not fortunate enough to see Harriet Martineau; she was from home: but we met several of our London friends wandering northward in search of the picturesque. It was pleasant to hear discussed the new theories of the season, and to laugh over the vagaries and superstitions which mark the steps of science, like the shadow of a man walking in the early sunshine.

It was delightful to hear my husband talking over with his friends these subjects of interest and amusement. The free air of the "pleasant land" where we sojourned infected his spirits with something of its vigour, and conversation dashed, foamed, sparkled like the little mountain torrents which made the landscape itself seem as if breaking into constant laughter.

We lingered on the shores of Winandermere and Ullswater; there were so many objects of interest to visit, so much to admire and enjoy. Mr. Crosse was deeply impressed by the beauty of the scenery; yet there was ever in his mind an upward, onward longing, which made even the grandest scenes and the loveliest combinations all-insufficient to fill up the picture with which the poet's imagination had environed his Ideal of Beauty. When he expressed this feeling to me, I reminded him that he should visit those parts of Europe where Nature has "ample room and verge enough" for displaying the magnificence of her architecture, and where she frames her landscapes in Alps and Apennines. "The same feeling would exist," he replied. "The mind of man carries him beyond the material beauty of this world, which is but the shadow of what his soul would grasp." Then, taking up his pen, he wrote rapidly some lines on "Change." I give the following extract, as it touches particularly on the feeling in question. My husband first describes all that fancy and reality can combine of natural beauty. He tells of rocks and cliffs that forbid the angry sea which foams at their feet; of mountains that rise grandly in their "cloud-capp'd" magnificence, forests and campagna, where the gorgeous vegetation of the East oppresses the sight, or where Italian skies smile kindly on the noblest works of art;—the varying cloud shadowed softly on the unruffled lake; the sun, rising in floods of crimson glory, distilling life in his beams; or the moonlight, calm and holy—and

> "Streams dancing in their sunny course, as if
> O'erjoy'd to pay their tribute to the deep;"

all this, and much more, he describes, and then he says:—

> "And yet, though glories such as these the wit
> Of man had ne'er conceived, had not they burst
> On his astonished sense, I feel a wish
> For things unseen—unknown,—some novelty
> On which my soul can rest, fatigued with scenes
> All earthly,—rocks, seas, stars and all
> That meets my vital sense: I want a sense
> Superior to that within, and freed

From worldly prejudice — with eyes to pierce
The unclouded wonders of my Maker's will,
And, knowing more, to bend more lowly. Yet
'T is not from discontent these yearnings rise;
Deep planted in my breast their seeds are found."

The early evenings of autumn began to close in before we left the lake district, now twice hallowed to me: it was my birth-place!

The winter of 1850-1 was passed at Broomfield in the quiet happiness of domestic life. Socially speaking, we were much isolated; for the Quantock range is at once picturesque and almost uninhabited, and Taunton, the nearest town, is six miles distant.

A person might walk from Broomfield, over the hills, for eleven miles, without even seeing a house or cottage, excepting dotted here and there in the distant vale.

I remember a long and very romantic walk which my husband took me this same winter. We went over the hills to Minehead, a distance of three and twenty miles. We walked the first day as far as Watchet, a little fishing town on the coast. Ere we reached it, night was coming on, and we lost our way in the intricacies of the numerous sheep-paths, which were our only guide over the hills. We have often laughed at the appearance we presented on the occasion of that memorable walk.

Mr. Crosse carried a telescope and a knapsack on his back, and a peculiar looking mineral hammer, with a long handle, in his hand, besides two or three other articles of philosophic apparatus. I carried a small basket of provisions, for we pic-nic'd just under Will's Neck, a point which commands one of the finest views in the West of England. It was here that Coleridge and Thelwall once sat watching the setting sun, when the former said, "Citizen John, this is a fine place to talk treason in."

"Nay, citizen Samuel," replied Thelwall; "it is rather a place to make a man forget that there is any necessity for treason."

We, too, talked of changes and revolutions, but they were such as science effects. The morning after we had slept at Watchet we started again on our travels, and greatly enjoyed the stroll by the alabaster rocks of Blue Anchor. The distance was nothing to my husband; for he had often walked the whole way from Broomfield to Minehead, literally, before breakfast, getting up at three, and arriving at his destination at nine o'clock.

When he was at home Mr. Crosse spent his mornings either in marking trees, or in superintending other work which, in an old rambling place like Fyne Court, was always demanding his time and attention. Economy was to us a necessary observance, too, in those days; and he was frequently engaged carrying out some new principle, applied to agriculture, mining, or manufactures, which my husband, in his sanguine enthusiasm, believed would be infinitely remunerative. In speaking of *knowledge*, some one, translating from the German, says: —

> "To some, she is the Goddess great;
> To some the milch-cow of the field,
> Whose business is to calculate
> The butter she will yield."

And science was worshipped by him in the elevated character of "goddess great;" at least the philosopher had none of the profits of the "milch-cow." A very *characteristic* American account of the Broomfield electrician says: "Crosse is a *retired* country gentleman, and has spent twelve thousand dollars on his apparatus," and goes on to remark that he is "living in a republican style of simplicity." I believe Mr. Crosse's scientific apparatus had cost some three or four thousand pounds. He used to say if he was immensely rich, he would build a national scientific institution, for educational purposes, and richly endow it: and he would add, that he would not accept of a large fortune if it were made conditional that he should keep a great retinue of servants, and live a fashionable life.

But to go on with the account of the way in which my husband passed his time. If the weather was bad, he would have fires lighted in the laboratory, and then there was plenty of work for every one. Batteries had to be taken to pieces or renewed; zinc rods to be cast, copper cylinders fitted for perhaps 150 bottles, which, with broken necks, stood up as battery jars. He was kind enough to make me his pupil, so that we had plenty of employment; indeed, the days were never long enough. In the evening he would read aloud, generally selecting history or poetry; but he used to say of the former, "I don't like history, for it is a tissue of the crimes and misfortunes of my fellow-creatures." Mr. Crosse read few scientific periodicals or works — too few, indeed; and thus often remained uninformed of what fellow-labourers were doing in the same field of research. Sometimes he would pass a morning in the workshop, busily occupied with his lathe; he was a first-rate turner, both in brass and ivory, and had made a good deal of his own apparatus.*

* Some one was observing one day, how beautifully he turned. "Yes," said Mr. Eagles, "Crosse can turn anything but — a penny!"

There was hardly any one near who cared for experimental science excepting my husband's second son Robert, who held the living of Broomfield, and was then residing at Kingston, about two miles from thence. His son and a chemical friend at Taunton, Mr. Draper, were almost the only people in the neighbourhood who were interested in his pursuits: he lost the society of the former early in 1852, when Mr. Robert Crosse removed to Ockham, in Surrey, to take possession of a living given him by Lord Lovelace.

Mr. Crosse frequently had friends from a distance staying with him, and on these occasions he was the life and soul of the party. His was a truly buoyant nature; he never was in bad spirits. He was the most uniformly joyous being I ever knew. Phrenologists said that the organ of "hope" was largely developed with him. Whatever the cause, he was certainly possessed of all the qualifications most delightful in a companion. He was at times much more like a schoolboy than a philosopher, and not only enjoyed, but was the promoter of, every species of fun and frolic.

Mr. Crosse spent some weeks in town in 1851. London will long remember the brilliant year of the Great Exhibition. It was impossible not to share in the intense excitement that then existed. The numerous scientific meetings, lectures, and *réunions* were particularly agreeable that season, from the influx of foreign *savans*. My husband enjoyed this kind of intercourse exceedingly. Some one very rightly observed of him, that he was "really a citizen of the world."

I shall not forget his first impressions of the Crystal Palace. We had spent the morning at the British Museum. There was hardly any one there besides ourselves, for the tide of sight-seers was gone westward. A strange fascination made us linger in the Egyptian rooms and in that part of the building devoted to the marbles of Nineveh. Mr. Crosse was always much impressed by things of antiquity. He stood gazing thoughtfully upon those religious and household relics which so strangely, so solemnly, unite us with the past. Time and humanity seem to have played a desperate game. Time was the victor, and these remnants of a bygone age are the trophies of the vanquished foe. Time lives on to conquer succeeding generations. All our "trophied arts and triumphs" are unavailing to stay the inevitable hand of the Destroyer. Proud of the possession of the present, do we believe that we can pass away, and our civilisation be "like a tale that is told?"

Something like this my husband said to me; and afterwards, when he paused in contemplation, I watched him: first, I saw an expression of pity for himself and for poor humanity, so short-lived, so feeble in possession; but there succeeded a gleam over his countenance which told of higher and holier thoughts, in which were mingled faith and hope — thoughts of *Immortality.* No one could say, truthfully, that he over-estimated human intellect: he often observed, "I believe the gate of heaven is kept by the angel of humility;" and few so truly realised the Preacher's admonition that "all is vanity;" and this in no narrow spirit.

Thus impressed with the thoughts of the past, he entered the Great Exhibition. Not all its gorgeous colouring, its dazzling light, nor even its rich treasures, fitly expressing the spirit of the age, could permanently remove from our minds the greater *reality* of the past.

After Mr. Crosse's return to quiet Broomfield, or at least somewhere about this time, the following letters were written.

"To William Spence, Esq.

"My dear Sir,

"It is now some time since you and I have exchanged communications, and my wife and I are desirous of knowing how you and Mrs. Spence are getting on. * * * A talented friend of mine has just written to ask me if you still retain the same opinion which you expressed in a pamphlet published by you about the year 1800, viz., "England prosperous without Commerce," with which he said he was greatly interested. For myself, I never knew you had written such a pamphlet. Will you have the kindness to let me know in your answer what I must say to my friend? He is an old school-fellow of mine, and lives at Clifton. We have had some highly accomplished friends staying with us: Walter Savage Landor was expected to join us, but was prevented by illness. There is but little society in these wild regions. I am working hard in my laboratory, and likewise very busy out of doors. * * * The Maker of the Universe seems to have established CHANGE as one of the first laws of Nature. Nothing sleeps; and the gold of California and Australia will establish two new and mighty empires to last, each for a time, and then to fall into the ruins of their predecessors. Have you seen the late account of *fresh* discovered cities in South America, supposed to have been destroyed long since by volcanic fires? It was a most interesting though unsatisfactory sketch in the *Times* a day or two since.

"My experiments are still progressing, and producing their *little* changes; a humble imitation of Nature—an endeavour to discover *Truth* without calling in the aid of human deception. My wife unites with me in every kind wish to yourself and Mrs. Spence. We hope we shall have the pleasure of seeing you both in this our wild residence.

"Believe me, my dear sir,

"Yours ever sincerely,

"ANDREW CROSSE.

"P. S. Kind regards to all London friends."

The visits which my husband alludes to our having received were from three school-fellows of his, and some other old friends who were with us at the same time. It was a delightful week; many circumstances combined to make it particularly happy and agreeable. One of our guests, a dear friend, now, alas! the *late* Reverend John Eagles, wrote a paper which was published in "Blackwood's Magazine," in January, 1853, in which he described the time he had spent at Broomfield. As it gives *his* impression of my husband's character, I venture to give some extracts. The paper is entitled, "Letter to Eusebius about Many Things."

"My dear Eusebius,

"I was sorry to hear of your accident, and should ere this have been with you, had not your cheering letter been put in my hands as I was making preparation to reach you. I therefore determined to prosecute my prior intention, and keep my promise to our old friend, and my old school-fellow, the philosopher, by visiting him at his ancestral residence amid the autumnal glories of the Quantock Hills. We conversed of you, and need I say how affectionately? * * * * * There is nothing more degrading to our nature than a low utilitarianism. And why, here, I throw out my indignation against those who would daub humanity over with the mud of their own thoughts, my Eusebius, will be apparent enough, when I give you some account of my excursion, my conversation with our poet-philosopher, amiably vehement — as a true, good, prejudiced man should be in all things,— and when I tell you of my seeings and doings after I left him. A prejudiced man! And would you admire a prejudiced man? will be the suggestion of the first common acquaintance who impertinently looks over your shoulder, Eusebius, while you read this. Yes, sir; I more than admire—I love a prejudiced man! * * * * You and I have not the same prejudices as our friend the philosopher, my delightful host, my old school-fellow; and both of us would be sorry indeed to uproot his, and graft our own in his stead. Prejudices make up identities. Without them we should be like only pease in a bag, and, like them, only fit for being boiled, and the worst of us thrown to fatten pigs. What are called 'strong-minded women' and men without prejudices are my abhorence; and having said this say, I look over my diary, and send you such extracts as may serve to amuse you. You are fairly down on your haunches, a ruminating animal:

take the nutritive grass out of my journal, and chew the cud at your leisure.

"*October—, 5 o'clock.*—Just come in from a walk with the philosopher and our mutual friend * ——; and before I dress for dinner, sit down to realise on paper this place and its improvements since I last saw it. It is a situation of singular retirement, amid the hills, yet at the head of a valley lengthening into some distance, sufficient for those various atmospheric perspectives which are the breath of beauty. Its character is pastoral. There is nothing dressed here, not even immediately about the house; but there are beautiful trees. The beeches prevail, whose silver stems so gracefully make a light in the deep wood shades. The large pond above the house has now an accessible path, where before there was a hedge; and as you ascend to it, the trees look very high and their large stems imposing. This is an improvement. I could wish the solitary swan had a companion. Poor bird he has lost his mate, and sails now gracefully up to greet every visitor. Philosopher should do as he would be done by; he is happily now no solitary bird; blessed be his nest! As we skirted the valley by the upland, the extent opened before us. The long hill-sides, heathery and of wood, not continuous, but with outstretching and receding patches, that slightly broke without destroying the unity, give a great air of a wild untouched freedom to the whole valley. As far as I have seen, these Quantock Hills have no large, barren,

* John Kenyon, Esq.

dreary, tableland, but are made up of slopes and dips; so that the moment you are at the top of one, you are close upon the descent into another.

"I have come to this conclusion, that even close to a house, in some situations, such as this, well-shorn grass lawn is not so pleasing to the eye as the ground covered with heather and fern, if beautiful trees grow among them; for how graceful is the fan-like fern! and there is a variety which the smooth-shorn lawn does not present. But in such a case there must be nothing trim; and I think also the house should be large as this is, and by its consequence show that beauty, not economy, is the object. Were I philosopher, I should be tempted to let the lawn ground be wilder. We see at this time of the year the advantage of this; for the red and orange-colour leaves strewed upon the ground, variegating the green, assimilate the ground below with the trees above, and take off the abruptness which is too visible where they are swept off. Nature loves not this abruptness, and strews the ground for a purpose. * * * It was a right pleasant meeting to-day when four old school-fellows sat down to dinner. How many years have passed since we were young and jocund in the same play-ground, and saddened over the same books! Our master was a good ripe scholar, and made scholars. I must pay a tribute to his memory, for though I left him earlier than my companions for a public school, he laid a foundation that I feel sure I have to thank him for of those literary tastes which have been my comfort and pleasure during a not very short life.

" One of the four I have not often met, but three of us, (and many of some mark, from the same school, I could mention) have cultivated literature ardently, and the philosopher the sciences also. Schools in our days were a little rough, it must be owned; perhaps in the end not the worse. * * * We four that met together to-day had had Orbitius's stamp upon the coin. We have no reason to complain of the interest it has paid us. I had a delightful walk this morning with the philosopher. Something or other touched a string that indignantly vibrated; and he broke forth in a strain of poetic satire that quite fascinated me. He must have repeated hundreds of his lines, and then, to undo the work of his wrath, he went off into continuous volubility of playful rhymes, then again to strains of tenderness. I did not think he possessed such a poetic vein, addicted as he has ever been to science. * * * We went to-day to see the church, which is almost within the grounds. There is an avenue of limes along the north side, and on the west and south are one or two very ancient yew trees, and a graceful old cross. The church is very old. * * * Every seat, excepting some modern ones, is exquisitely carved, in a great variety of patterns and monograms. * * * Generations who have occupied those seats are but dust; and the dry dust, like the colour of the wood, seems given by hands that, in the solitude of midnight, have come out of their graves and gone fondly over the monograms and devices of their race.

" I left yesterday the agreeable hospitalities of my friend the philosopher." * * * * *

Mr. Eagles has so well described the *locale* of Broomfield, that I have been tempted to extract more at length from his pleasant paper than I had at first intended. It was very delightful to see my husband and Mr. Eagles together: they differed widely on some points, but they were "closely united in friendship." Some letters addressed by Mr. Crosse to Mr. Eagles during the years 1852-3-4, are here subjoined.

"My dear Eagles,

"What is a man to do who is on the point of bursting with indignation? I answer, 'Find out one who is capable of feeling a similar indignation, and halve it with him.' Now, Eagles, thou noun of multitude, I turn to thee. I must first premise that there are sundry points, and those of no slight importance, in which you and I differ — but we both of us have a *reason why* we differ. It is a difference arising, not from want of reflection on either side, but from a different mode of arriving at our conclusions: each of us, feeling that we are free agents and slaves to none, forms his judgments as he best can, and gives reins to his fancy, which, however, is cultivated, more or less, by reference to those *great men* of all descriptions, — poets, painters, musicians, historians, and others, — whose names will live when this nameless age shall have passed away.

"In this, I presume, we coincide. In the present age they *hammer*, they *file*, they make *steam engines*, and a few philosophers, but more materially than mentally. Sublimity is DEAD. Tennyson's Ode to Wellington is called poetry—Milton, Virgil, and Pope are called no poets—and the other day I heard the editor of a London newspaper say that Horace's Odes contained *no poetry!* * * * My wife and I have just spent three weeks in town: when at the Exhibition of pictures, we heard the most extravagant praises of all that least resembled nature, and the greatest abuse lavished on all that had the misfortune to approach it. * * * While in London my wife and I had a long talk with Faraday, and with other men of eminence, and their opinion is that in spite of the outcry about advancing civilisation, it is only the few, and not the many, who are really civilised. In this I cordially agree: I once said sportively in verse—

"'Sure Nature works by unfair spell,
And steals men's wits to heap on one;
And though she forms a miracle,
More than an empire are undone.
Thus is produced by certain rules
One wise to many thousand fools.'

"I do not arrive at this conclusion without sundry misgivings. If the whole genus are sunk so low, how must one look down upon ONESELF? I must confess, however, that my opinion of the intellect of my fellow creatures is much lowered within the last few years.—What with spirit-rapping, mesmerisms, table-turning, &c.—an acceptation of all that is absurd and vile, and a rejection of all that is sublime and noble—what can I think? Alas! a proper education seems farther off than ever, and how few are qualified to instruct! My wife, I am sorry to say, is by no means well. I need not say how happy we shall be to see you and Mrs. Eagles.

"We both unite in every kind wish, and believe me, my dear Eagles,

"Yours ever sincerely,
"ANDREW CROSSE."

"My dear Eagles,

"Your letter lies before me in which you attack ——. He evidently does not comprehend Swift, and looks back upon him, through his own distorting lens. How truly do you say, 'How little do we know of characters; and how should we? Every character is an individual mystery, and yet how meagre is our nomenclature. We can scarcely describe what any man is.' My good, kind, metaphysical brother used to say, 'our principal fault is the not making sufficient allowance for the conditions under which our neighbour forms his judgments.' People are divided into two classes, the worldly and the unworldly. The first I *hate*. I do not quarrel with them, but treat them as children *without their innocence*. The last I love, and you, my good friend, are of the latter. By the bye, how is Kenyon? Have you heard of, or from him lately? Do tell me in answer. * * Our little boy is grown a fine fellow, and his intellects are rapidly being developed, and his love of mischief getting stronger daily. My own health is surprisingly good, far better than I expected, or deserved. I have indeed *much* to thank my Maker for, who has raised up to me in my latter days (after a life of repeated trials, such as few have been called on to suffer) the kindest, most affectionate, and truest friend I have ever met with. * * * I have always felt that one atom of real kindness is beyond all the frippery and parade of the selfish part of mankind * * * This brings me to a point in your letter. Is a mercantile community more advantageous than an agricultural? Is it more virtuous? more self-denying? I confess that I should be loth to rely on the opinion of the wisest in such a matter. There is scarcely one, if any, man breathing who would not answer according to his *prejudices* to such a question. Maritime Athens fell before the power of agricultural Sparta, and commercial Carthage before military Rome. It should seem, I *fear*, that commercial nations are more short-lived than others. My wife and I have lately been reading Gibbon, and I was much struck with the resemblance between the present state of France and that of the Roman Empire a little previous to its final destruction. Chemically speaking, the constant breathers of oxygen gas would have an *excited* but short life. Is not this somewhat the case with a commercial nation? However, to judge from my own experience, limited as to the commercial, but more extended in the agricultural view, I have *little* or nothing to say in favour of my own class. * * * * I believe the greatest amount of practical virtue exists amongst the poor mechanics, in whom I have met some magnificent traits of character. No improvement worth mentioning has, I believe, been derived to the country from the landed interest, except that of furnishing the necessary food for man. They have invariably resisted all change for the better, and supported unnecessary, *unjust*, and ruinous wars. I have written to my excellent friend Spence, and asked him the question you desire. My wife and I shall attend an evening meeting of the Archæological Society, to be held at Taunton to-morrow, after we have dined with the Kinglakes. * * * *

"Yours ever sincerely,
"ANDREW CROSSE."

"My dear Eagles,

"Mrs. Robert Crosse, my son's wife, last night handed me your very kind and affectionate little note, which I am delighted to answer in the same spirit. I have not written any letter to you; and had I done so it would have been worded simply in the manner of inquiry, and most certainly not in that of blame; as I am not one to judge without hearing both sides, and, moreover, I could not suppose for one moment that you would act unfairly by me. Disputes of all sorts I *abhor*, and to say the truth I have often allowed myself to suffer most unfair treatment by *soi-disant* scientific men, and who ought to have known better, rather than enter into tiresome discussions with them. Besides, I am an awful unbeliever, and most of what enters one ear slips out of the other. I cannot describe how much your communication has pleased and affected me. You and I have known each other too long to quarrel; and friends such as we are are rarely to be met. * * *

"Banish, therefore, from your mind every idea of my being otherwise than your *firm* and devoted friend,

"ANDREW CROSSE."

" To Dr. Brittan.

" Broomfield, near Taunton,
" May 7th, 1854.

" My dear Dr. Brittan,

" Some years ago I wrote some lines on Poland, which being particularly applicable to the present time, I forward you a copy of, having lately had a few printed for my friends. I am a warm lover of *peace*, but not such peace as Bright and Cobden would give us — purchased at the expense of our *national honour*, and a base and venal submission to blood-thirsty and ambitious tyrants. The present war is a holy one, and a fight for the dearest interests of society against a crafty, merciless ruffian. May God defend the cause of justice, and bring the present sad, but necessary, war to a speedy and successful termination !

" We are all well here, my wife much better than she has been for many years.

" This is my favourite month, when Nature puts on her glorious attire, and each fruit-tree vies with its neighbour in profuse rivalry of blossom in honour of its great Maker. What a contrast to the stormy passions of selfish man ! Yet this world is, as it was designed to be — not a heaven nor a hell — but a mixture of *both*, a contrast for its dwellers to select from, at least to a certain extent. It is the contrast which causes the two opposites to heighten each other, and by a beautiful admixture of contending principles to purify our souls, soften our asperities, and refine us for a state in which evil is unknown, and good *triumphant :* without this hope, what would this strange assemblage of worldly passions be? Even science with its allurements, and literature with its refinements, would not suffice. We pant for knowledge without its paltry jealousies, and disencumbered of the veil which will ever obscure it in this variable dream. We have that within us which may be worthy a better habitation, and which we trust it will have when it leaves its dregs behind. Is not the mind the human temple of the Creator, from which, as far as human power is able, especially if aided by Divine influence, the baser passions should be expelled? I am inclined to smile at my own unintentional sermon: you will forgive me this tirade, which I have rapidly sketched as it were in spite of myself.

"My wife unites with me in every kind wish to you and Mrs. Brittan, and

"Believe me,

"Yours very sincerely,

"ANDREW CROSSE.'

"Clifton, September 28th.

"My dear Miss Douglas,

"* * I don't envy you the difficulty of decyphering this labyrinthian piece of penmanship, but it must take its chance. I am well aware that it is unlike any other scrawl — but then I was born a rebel to all *past* and *present* conventionalisms. Then to believe in them, — I might be a *Mesmerist*, a *table-turner*, a *rappist*, a *Mormonite*, or a thousand other atrocities which ride triumphant over the senses of the multitude,—to say nothing of the streams of light which those possessed of far sharper senses than I have can see issuing from the poles of a magnet in the dark. Such are the effects of 'civilisation,' as it is called: truly this is a strange world. Nature would seem affronted by the destruction of the original *chaos*, and to have planted it in the mind of man. Cornelia and I left our quiet home on Saturday last, to visit our kind friends the ——. Theresa is to return to us on Saturday, after a long series of visits. * * Our eldest babe is growing a fine fellow, with all his mother's love of mischief; he runs about like a lamplighter. He is the delight of his grandmother, to the discomfiture of Theresa, who is a jealous animal. I have of late been engaged in following up closely my favourite science, and have been highly successful, in the 'dropping' experiment near the laboratory, which was set in action when you were with us. I have crystals of arragonite growing at the negative end of the piece of clay-slate in some quantity; and, also, what I am almost certain are crystals of quartz growing in considerable numbers on the positive end. I have also formed an entirely new mineral in brilliant *octohedral crystals*, now forming upon a coil of platinum wire. These crystals are composed of *oxygen*, *silver*, and *copper*, and such are not known in *Nature*, nor have they hitherto been formed by *Art*. My wife very ingeniously hit upon a compound voltaic battery which brought these crystals to light. Besides this I have discovered a method of reducing gold to powder by the mechanical action of carbonic acid gas, elicited by electrical action. I mentioned this to Faraday some time since, and he was much pleased with it. You must come to Broomfield, and see with your own eyes these arrangements, which are making faint efforts to dive into the mysteries of Nature — or rather the glorious works of the Omnipotent.

"Cornelia and I have been a second time to Battleborough, to prepare for the erection in the ensuing spring of some new buildings necessary for the farm. We are much pleased with the situation, which is indeed most beautiful, and which commands one of the finest views in our beautiful county. At Broomfield we are cutting down a large quantity of laurel and under cover round our house, to give a better view of the large trees, and admit the air. We have lately built a new cow-house, and taken in hand twenty-five more acres, so that Cornelia is becoming a *she* farmer. *I* have nothing to do with her farming. A new school house is now building at Broomfield, near the church; and the singing gallery is taken down in the church, and considerable improvements about to take place. * * * Cornelia sends her best love.

"Believe me, my dear Miss Douglas,

"Yours very sincerely,

"ANDREW CROSSE."

About this time, while staying with his relative, Henry Porter, Esq., of Winslade House, near Exeter, Mr. Crosse gave some lectures on science, at the request of different local literary and mechanics' institutes in the south of Devon; but being intended for a mixed audience, they comprised mostly a repetition of some of his early experiments.

In the large and pleasant circle at Winslade he was the life of the party. The younger portion of the community would seize upon him, asking him to tell them humorous stories, of which he had a great stock. The merriest response was always ready to reward the narrator at the right moment. Mr. Crosse and his cousins too "came of a *laughing family*," as one of their little boys said, by way of excusing himself for an indecorum at school.

During those hospitable joyous days, I am sure we were ready, all of us, scientific and unscientific, to subscribe to the wisdom of what a distinguished chemical philosopher once said to me, namely, "that if he had a good hearty laugh, he never did, or felt, a wrong thing through the day." Most admirably does the author of "Companions of my Solitude" touch upon the reverse of this, when he says, "there is a secret belief amongst some men that God is displeased with man's happiness, and in consequence they slink about creation, ashamed and afraid to enjoy anything." Mr. Crosse was certainly not one of those who would wish nature to put on a quiet "suit of drab," instead of her polychromatic chorus of joy.

In the autumn of 1854, Mr. Crosse attended the meeting of the British Association at Liverpool. We were the guests of Joseph Yates, Esq., of West Dingle Park, where we found assembled a distinguished party of scientific persons; amongst them the great geologist, Sir Roderick Murchison, and William Hopkins, Esq., President of the British Association for the year 1853. That gentleman, as is usual, opened the proceedings of '54, resigning the chair to the new president, the Earl of Harrowby.

My husband had never been at any of the meetings of the British Association since his first appearance, several years ago, at Bristol, and now he took his place merely as a looker-on. He mentioned incidentally to Sir Roderick Murchison the experiment elsewhere detailed, whereby he discovered a very singular power in *nascent* carbonic acid gas. He found that a sovereign placed on a piece of marble, with the positive terminal of a voltaic battery put immediately between it and the coin, the same being kept in place by a glass weight, and the whole being submerged in a *weak* solution of sulphuric acid and water, connected with the negative pole, in the liberation of carbonic acid gas from the electrically decomposed marble or carbonate of lime, it had the power of breaking the gold coin to pieces, apparently by mechanical action.

Sir Roderick was much struck by Mr. Crosse's account of this experiment, and observed, "that if the subject were pursued to its conclusions, it might go far to account for much unexplained phenomena in nature." That eminent geologist advised him to write the full details of the experiment. Mr. Crosse did so, and read his paper at the Chemical Section.

Professor Miller, the president, and two or three others, objected to the conclusions drawn by Mr. Crosse. They considered the effect due to the disintegration of the copper, by the action of the dilute sulphuric acid. The gold acted on, it must be remembered, was a common sovereign, of course containing copper. The scientific men then present maintained the opinion that if *pure* gold had been used, the same result would not have ensued. Mr. Crosse frankly acknowledged the reasonableness of the objection, stated his readiness to try the experiment in the way suggested; but he gave it as his opinion that the carbonic acid gas, at the moment of liberation, would be found to have the same power in respect to chemically pure gold as it had in the case of the sovereign. A dark stain on the white marble (which was produced at the Section) was by Professor Miller attributed to the presence of copper. Mr. Crosse considered it to be the purple oxide of gold, because, he observed, had the sulphuric acid attacked the copper, it must have been found at the negative pole of the battery, which was not the case.

Owing to a combination of circumstances, it was some months before my husband again tried the experiment, and I will only here briefly observe that the result of the new trial was *perfectly* successful, and fully confirmed the views which he brought forward on this occasion.*

This meeting of the British Association afforded Mr. Crosse an opportunity of renewing his intimacies with many scientific persons, whom he had not seen for a long time; some whom we had not known before became our sincere friends. My husband often reverted with real pleasure to the cordiality which existed amongst the strangers collected together at this time, and above all to the hospitality, public and private, which marked their reception at Liverpool. We were not personally acquainted with our kind host (now, alas! no more) and his daughter till we

* The details of this experiment "On the apparent Mechanical Action accompanying Electric Transfers" are to be found in Chap. IV.

became their guests; but Mr. Crosse had long known his brother, Mr. James Yates, of Lauderdale House, Highgate — well known for the prominent part he has taken in advancing the decimal system.

The wealth of the Liverpool merchants is certainly most nobly expended in the cultivation and advancement of the fine arts. The occasion of this brilliant meeting gave an opportunity for the display of the emulative patronage which artists and architects had received at the hands of newly created wealth. My husband was much struck by the estimation in which science and literature were held in circles whose tone of thought was new to him, coming from the far west, where civilisation has unhappily stuck in the mud for the last two centuries. But there is no light without its shadow, no good without its accompanying evil. I remember my husband repeating to me a conversation he had had with one of the "merchant princes" of this modern Venice. "We have *too much* money," observed the gentleman in question; "the possession of it has altogether an undue preponderance with us: it is our universal gauge." Then he went on to say, "To give you *our* definition of the word '*good*' will explain what I mean. By a 'good man' simple hearted people mean a moral, virtuous man; at the universities I understand it means a first rate scholar; with us 'a good man' is a man of capital, a safe debtor!!"

St. George's Hall, one of the most magnificent, perhaps the most magnificent public room in Great Britain, was opened at the meeting of the Association. We were present at the oratorio of the "Creation" on Wednesday, September 20th. And while the deep thunder of the splendid organ, containing eight thousand pipes, reverberated through the building — we knew it not then, but other peals of thunder were echoing from the heights of Alma; our heroic countrymen were dying and conquering. It has often seemed to me strange how little in reality the general pulse of society is quickened by events which, read afterwards in history, appear the all-absorbing life of the day.

"Broomfield, near Bridgwater,
"Sept. 30th, 1854.

"My dear Theresa,

"Thanks for your letter, which I now reply to. In the first place you may wish to know what has taken place here since your absence. Our little Babo was, you know, thin and puny; he is now grown strong and fat, and quite well. Big Babo is quite well, stout, and happy. * * *

"On Monday the 11th Mr. and Mrs. Henry Philipps and their niece arrived here. On Tuesday morning we started for Taunton, to be present at the anniversary meeting of the Somerset Archæological and Natural History Society. We all met at the Assembly Room, Mr. Labouchere, the member for Taunton, being in the chair. Four different papers were read, some of them interesting. After this we adjourned to the London Inn, where there was a public dinner. Cornelia was led into the room by Mr. Labouchere, who was President, and all our party sat on his right hand, and did justice to a capital dinner. In the evening there was another meeting; I read a paper with some difficulty, for the room was so scantily lighted. The next morning the Bailiffs of the town of Taunton gave a public breakfast; but we could not attend, as we had invited the whole party to a one o'clock luncheon at our own house. Accordingly, a little before that hour, more than fifty carriages drove up to Broomfield Green; and after inspecting the church, the whole party, consisting of more than two hundred persons came down to Fyne Court. * * * I had prepared some experiments in the electrical room, and covered my electrical table in the music room, with a variety of choice minerals from the Quantock Hills. All these matters called forth great curiosity, and I was half killed with answering questions. Mr. Calcott, son of the eminent composer, played on the organ, and the whole was the gayest scene possible. People from very distant parts of the kingdom joined our party. We had not had such a gathering since the inauguration meeting, when poor Dr. Buckland was present — almost his last appearance in public. * * * * * * *

"On the 19th we went to Liverpool, to be present at the meeting of the British Association. I can give you no adequate idea of the parties we have gone to — dinners, soirées, &c., concerts, and all sorts of festivities. I was at a dinner party at the Mayor's, which went off very well. Thirty-five sat down. We attended a soirée at the Town Hall. There was a perfect cram, and fifty thousand pounds' worth of the finest modern paintings were exhibited, which were lighted up in the most tasteful and well-judged manner. We received every kindness and attention from every one. * * * I read a paper on my favourite science at the Chemical Section, and exhibited some specimens of electrically-formed crystals. Mr. Hopkins, the late President, and Sir Roderick Murchison were staying in the same house with us. Quiet Broomfield, with its silent trees and uninhabited hills, is a great but not unpleasant contrast to all this.

* * * * * * * * *

"ANDREW CROSSE."

A friend reminded me lately of a slight circumstance which took place during the archæological party at Fyne Court. Mr. Crosse was eloquently explaining to a group of the most scientific of his guests some of the higher branches of electricity, when a gentleman, who had just joined the circle, interrupted him with, "Will you explain to this lady the difference between positive and negative?" He turned round, I understand, with the greatest good humour, and gave the required information.

The autumn and winter of 1854 were spent quietly at Broomfield. A cloud of recollections comes over me as connected with those well remembered months, but the *incidents* are few and of *no* significance, except perhaps the *deepest*. Within the sacred circle of *home* man lives the truest, both to himself and others; but of "noted action" in "this daily history" what is there that the world cares to hear? The record of the happiest of lives was simply these monumental words, "I, too, was an Arcadian."

I have often wondered, if a minute and faithful transcript of any life could be given, whether it would weary us with its prose or astonish us with its poetry. As for the good and evil in life, Mr. Crosse believed them to be equally balanced, at least in the world; I do not mean to say with respect to individuals. He was once present at a small party, where the question was put to the vote as to whether there was an excess of good or evil. Fifteen believed in the predominance of *good;* fifteen thought there was a greater amount of *evil.* Mr. Crosse had the casting vote, but he said, "I will not disturb your verdict, for I believe that there is just as much good as evil, just as much evil as good — all tending towards future good."

The subjoined letters are almost the only chronicles I can find heart to give of happy days, now briefly numbered.

"Broomfield, February 16th, 1855.

"My dear ——,

"The weather has been the most awful I ever witnessed. Every large tree was laden with tons of ice, and when the thaw took place there was an incessant rain of icicles, which brought down with them the boughs and limbs of some of our finest trees; literally, they came 'cracking, crashing, thundering down,' doing incalculable mischief to the woods and plantations. I never saw so universal a destruction of timber. The season has been very remarkable. The supply of water has been unusually scanty. The leaves had remained longer on the trees than I ever knew them. The oaks, only one month before Christmas, had all their leaves on, and were not even touched by the autumnal tint.

"To-day, with a large fire in the music room, the thermometer is two degrees and a half below freezing. In endeavouring to open the lid of the tea-pot, we found it frozen hard to the pot, with a rim of ice all round it! This in front of a large wood fire. The wind is roaring awfully, and the windows are rattling in their frames. My electrical glasses are exposed to imminent danger from their fluid contents being turned into ice. I am obliged to have fires in all directions. The snow on Broomfield Hill is drifted in some places to six feet deep or more. Last night the thermometer out of doors was thirteen degrees below freezing, and this before midnight. * * * *

"Yours, &c.

"ANDREW CROSSE."

In April, 1855, Mr. Crosse went to London. While there he had the pleasure of seeing a great deal of his dear friend and old schoolfellow, Mr. Kenyon. The evening before he left town he dined at his house, in company with Mr. Eagles: the three friends never met again! The tone of the following letter does not indicate any impression of the approaching change.

"36. Russell Square, London,
"April 28th, 1855.

"My dear Miss Kinglake,

"Your letter followed me to Russell Square, where my wife and I are now staying. We return home about the middle of May, when our kind friends the Coxes will pay us a visit at Broomfield.

"Your communication is full (I won't say of error, as ladies of course never err, but) of mistakes. Our *kind, good* friend Kenyon is *not* the person referred to as 'Malignant,' &c., in the 'Memoirs of Lady Blessington;' but it is, I hear, some defunct Frenchman. The story is totally at variance with Kenyon's *well known* character, and I wish you would be kind enough to contradict it. We intend to have it contradicted in print.

"We have a peck full of congratulations on the birth of our last son, which took place on March 23rd. * * As for my old electrical pursuits, I am more engaged in them than ever, and with success. At present I don't anticipate attending the Glasgow meeting. We are very busy here. Last night we attended a geological lecture of Sir Charles Lyell's at the Royal Institution; our friend Faraday supplied us with tickets. We take tea this evening with poor Spence, who has lately lost his wife.

"The universal feeling of all people in town and country is in favour of putting down all aristocratic influence, and getting our affairs managed by *another* class of people. Even in this case it does not follow that matters will be better managed. * * * * * * I can write freely to you, as I know you to be a rebel, and 'fit for treasons, stratagems, and spoils.' I find that I have undergone two alterations of late: *one*, that there is vastly more real kindness and affection in the world than I used to believe; witness the kind, feeling, generous, unselfish letters from our common soldiers in the Crimea. Secondly, that the masses, high, middling, and low, have far less *judgment* than I could have expected in so reasoning an animal as man is falsely called. Look at the immense variety of *sects* of every denomination, &c. &c.; and far beyond all other folly, the gross servility of the many to the few. Were it not for this, war would be expunged from the earth; the first tyrant that arose would be cut off: but we cannot alter matters, and must make the best of an imperfect world.

Excuse this diabolical scrawl, and with my wife's kind regards,

"Believe me,

"My dear Miss Kinglake,

"Yours ever sincerely,

"ANDREW CROSSE."

"36. Russell Square, London,
"May 11th, 1855.

"My dear Theresa,

"Your sister is much fatigued, I therefore write to give directions about our carriage meeting us at Bridgwater Station. We return on Tuesday. * * * * Yesterday morning we breakfasted with Kenyon, and met Crabbe, Robinson, a son of Wordsworth, a brother of Southey, and others. To-morrow we dine with Kenyon, and meet Eagles. On Wednesday we went to the Botanical Fête, which was largely attended, and the flowers magnificent beyond description, such as I never saw before anywhere. The weather was horridly cold. We hope to have a letter from you to-morrow, bringing accounts of the three dear children. * * Since we have been here we have of course met persons of every shade of opinion, but they one and all agree that our government have shown a general want of order and arrangement. This accounts for the horrible catastrophe at Sebastopol; the sacrifice of so many thousands of brave men. * * * * * * *

"Believe me,

"Yours ever sincerely,

"ANDREW CROSSE."

We returned to Broomfield on the 15th of May. We had some friends staying in the house, and they remarked that they never saw my husband in better health or higher spirits. On the 22nd we had occasion to go to a distant farm on business, and we walked that day seven or eight miles, and returned without being in the least fatigued.

On the 23rd he passed the whole morning in the laboratory, and set in action a very interesting experiment; he pursued these arrangements with his accustomed ardour. The subject of investigation was the same which he had brought forward at the British Association meeting at Liverpool.

Mr. Crosse had previously arranged twelve jars of Daniel's sustaining battery, which when charged deviated (if I recollect rightly) a single wire galvanometer forty-five degrees. He then took a large glass jar, filled it with dilute sulphuric acid and distilled water, and placed at the bottom of the vessel a small square of white marble, on it a solid piece (coin shaped) of *chemically pure* gold. By an arrangement *, the details of which I need not repeat here, he caused the carbonic acid gas, at the moment of ex-

* The details are given in Chap. IV. See the paper contributed by Mrs. Andrew Crosse on this subject to the meeting of the Association in 1855.

trication, to knock off, apparently by mechanical action, no less than twenty-three grains from the gold. This result was most conclusive. In the former experiments, with a less powerful battery, he had succeeded in breaking off six grains. The objection raised, that the result was owing to the disentegration of copper in the sovereign, was now completely answered, as the gold now used had been prepared for the purpose, and was chemically pure.

My husband was greatly pleased at the success of the experiment, which was certain before the lapse of twenty-four hours; but he did not take it apart, for he resolved to leave it at work as long as the battery continued strong enough to cause a powerful evolution of gas.

This was the last scientific act of his life. He did not live to take abroad this experiment, so ingeniously, so beautifully arranged; and when, during his illness, I spoke of it, he more than once stopped me, saying, "Don't talk to me about it *now*," though on other matters of science he would discourse freely, almost till the last.

On Friday, the 25th, Mr. Crosse gave his annual rook-shooting dinner to some of his tradespeople and tenants. He joined them in their sport, and seemed particularly cheerful. Later in the evening, he took a walk alone with me, when he seemed somewhat depressed, and complained of a sensation of heaviness in the legs; but it passed off, and we took no more notice of the circumstance.

The next morning, the 26th, while dressing, he suddenly complained of giddiness. For the first few moments I was not much alarmed, thinking it was merely a nervous attack to which he was occasionally subject. He threw himself on the bed, and immediately afterwards he felt a deadness on the left side; he became very pale, and turning to me he said, "Cornelia, I have a paralytic seizure; send for Mr. King."

At first, I could not, I would not believe it; but, alas! his words were only too true. I tried to attribute the sensation he was suffering to any rather than the right cause; but he was more used to sickness than I had been, and he knew the fatal truth.

With deep emotion, yet with calmness, he said, "My dear wife, bear it, as I must bear it. Do not deceive yourself, this is my death-stroke."

During the three agonising hours that intervened before the medical man could reach Broomfield, my dear husband pronounced these solemn words, which I shall never forget — "If by moving my finger I could restore myself to perfect health, and to the certainty of several years of life, I would not do it if I knew it to be contrary to the will of God." This sentiment uttered at such a moment of trial, struck down as he was in the midst of strength and happiness — by the awful summons of death — is indeed memorable.

* * * * *

* * * * *

The closing scenes of life are almost too painful to dwell upon, yet we must not shrink from the glorious example afforded by a humble and sincere man, trusting in faith to the mercy of his God, and passing from this chequered existence with the firm hope of another and a better world.

While memory lasts, I shall never forget the solemn nights of watching that I passed, often alone, with my beloved husband. He rarely slept, and through the dark hours of night, or during the early dawn of those bright summer mornings, he would talk to me on subjects of the highest and holiest interest. At times his mind seemed intensified by his illness, his conversation was such that it was a strain upon my own mental powers to follow him. I could almost say that he anticipated and realised the greatness of immortality. But then, again, at other times, physical weakness would lay him low, and the powers of life seemed ebbing fast; he would then often be nervously depressed and anxious for the visits of Dr. King, whose kind attentions always soothed him.

Latterly he suffered intense pain, but bore it with the utmost patience. I never heard a complaint or a murmur pass his lips. He said, more than once, "I have no reason to complain, I have lived my time;" and he would observe, "Morning and evening seem all the same to me now, but I pray constantly."

One day when he felt somewhat better, he repeated to his eldest son and myself his translation of Horace's Ode to Augustus Cæsar; and about a week before his decease, in allusion to his own state, he quoted his own lines "To the Chamber Clock."

His beloved friend Mr. Eagles spent some last sad days with him. His first family also attended the summons which called them to his death-bed. On Sunday, July 1st, he received the sacrament, at the hands of the clergyman of the parish.

After this time he became rapidly worse; the symptoms he suffered from were more distressing and painful. Two days before his death, he sent for our three little boys, and blessing them, he said, " I do not pray that they may be great, or that they may be rich, but I pray that they may be *good*."

On the morning of July 6th, my beloved husband breathed his last.

* * * * *

* * * * *

Andrew Crosse died in the room in which he was born, beneath the roof where he had lived; and his mortal remains rest where his ancestors have been laid for more than two centuries.

On the simple monumental obelisk which mark the place of his last home are these words:—

SACRED TO THE MEMORY OF

ANDREW CROSSE,

THE ELECTRICIAN,

BORN JUNE 17TH, 1784. DIED JULY 6TH, 1855.

He was humble towards God and kind to his fellow-creatures.

THIS TRIBUTE OF AFFECTION IS RAISED BY CORNELIA, HIS WIFE.

NOTE

ON THE ACARUS ELECTRICUS.

In the year 1837 Mr. Crosse was pursuing some experiments on electro-crystallisation, and in the course of these investigations insects made their appearance under conditions usually fatal to animal life. Mr. Crosse never did more than state the fact of these appearances, which were totally unexpected by him, and in respect to which he had never put forth any theory.

What he thought of the power of electricity is best expressed in his own words. In speaking of his favourite science, he says: —

"Electricity is no longer the paltry confined science which it was once fancied to be, making its appearance only from the friction of glass or wax, employed in childish purposes, serving as a trick for a schoolboy or a nostrum for the quack; but it is even now, though in its infancy, proved to be connected most intimately with all operations in chemistry,—with magnetism, with light and caloric, apparently a property belonging to all matter, and perhaps ranging through all space, from sun to sun, from planet to planet, and not improbably the secondary cause of every change in the animal, mineral, vegetable and gaseous systems."

In especial reference to these experiments in which animal life appeared, he says: —

"I have met with so much virulence and abuse, so much calumny and misrepresentation, in consequence of these experiments, that it seems, in this nineteenth century, as if it were a crime to have made them. For the sake of truth and the science which I follow, I must state that I am neither an atheist, nor a materialist, nor a self-imagined creator, but a humble and lowly reverencer of that Great Being of whose laws my accusers seem to have lost sight. It is my opinion that science is only valuable when employed as a means to a greater end. I attach no particular value to any experiments that I have made, and I care not if what I have done be entirely overthrown, if Truth is elicited. Though warmly attached to experimental philosophy, I have never for one moment imagined that it is possible to perform a single experiment which is absolutely perfect in itself, or indeed that we can carry out any train of such which are not more or less liable to objection."

In speaking of the misapprehension of prejudiced persons, he says: —

"By such I have been termed a self-imagined Creator. Man can neither create nor annihilate. To create is to form a something out of a nothing; to annihilate is to reduce that something to a nothing. The chemist plays with the substances brought under his notice; he decomposes; he recomposes; he is a humble imitator of Nature; to create or annihilate is not in his power."

I now give Mr. Crosse's own account of the first experiment in which the acari made their appearance: —

"In the course of my endeavours to form artificial minerals by a long-continued electric action on fluids holding in solution such substances as were necessary to my purpose, I had recourse to every variety of contrivance that I could think of; amongst others I constructed a wooden frame, which supported a Wedgewood funnel, within which rested a quart basin on a circular piece of mahogany. When this basin was filled with a fluid, a strip of flannel wetted with the same was suspended over the side of the basin and inside the funnel, which, acting as a syphon, conveyed the fluid out of the basin through the funnel in successive drops: these drops fell into a smaller funnel of glass placed beneath the other, and which contained a piece of somewhat porous red oxide of iron from Vesuvius. This stone was kept constantly electrified by means of two platina wires on each side of it, connected with the poles of a voltaic battery of ten pairs of five-inch zinc and copper plates. The droppings of the second funnel fell into a wide-mouthed bottle; and they were poured back again into the basin, when that vessel was getting empty. It must not be supposed that the stone from Vesuvius was in any way connected with the result of the experiment. It had been selected principally for its porosity. The fluid with which the basin was filled was made as follows: A piece of black flint, which had been exposed to a red heat, was reduced to powder. Of this powder two ounces were taken, and mixed intimately with six ounces of carbonate of potassa, and then exposed to a strong heat for fifteen minutes. The fused compound was then poured into a black lead crucible in an air furnace; it was reduced to powder while still warm; boiling water was poured on it, and it was kept boiling for some minutes. The greater part of the soluble glass thus formed was taken up by the water. To a portion of the silicate of potassa thus formed I added some boiling water to dilute it, and then slowly added hydrochloric acid to supersaturation.

"The object of subjecting this fluid to a long continued electric action through the intervention of a porous stone was to form if possible crystals of silica; but this failed. On the fourteenth day from the commencement of this experiment I observed through a lens a few small whitish excrescences or nipples, projecting from about the middle of the electrified stone. On the eighteenth day these projections enlarged, and struck out seven or eight filaments, each of them longer than the hemisphere on which they grew. On the twenty-sixth day these appearances assumed the form of a *perfect insect*, standing erect on a few bristles which formed its tail. Till this period I had no notion that these appearances were other than an incipient mineral formation. On the twenty-eighth day these little creatures moved their legs. I must now say that I was not a little astonished. After a few days they detached themselves from the stone, and moved about at pleasure.

"In the course of a few weeks about a hundred of them made their appearance on the stone. I examined them with a microscope, and observed that the smaller ones appeared to have only six legs, the larger ones eight. These insects are pronounced to be of the genus *acarus;* but there appears to be a difference of opinion as to whether they are a known species; some assert that they are not. I have never ventured an opinion on the cause of their birth, and for a very good reason — I was unable to form one. The simplest solution of the problem which occurred to me was that they arose from ova deposited by insects floating in the atmosphere and hatched by electric action. Still I could not imagine that an ovum could shoot out filaments, or that these filaments could become bristles, and moreover I could not detect, on the closest examination, the remains of a shell.

"Again, we have no right to assume that electric action is necessary to vitality until such fact shall have been most distinctly proved. I next imagined, as others have done, that they might have originated from the water, and consequently made a close examination of numbers of vessels filled with the same fluid: in none of these could I perceive a trace of an insect, nor could I see any in any other part of the room.

In another experiment Mr. Crosse observes: "I used a battery of twenty pairs, between the poles of which were interposed a series of seven glass cylinders, filled with the following concentrated solutions: — 1. Nitrate of copper. 2. Carbonate of potassa. 3. Sulphate of copper. 4. Green sulphate of iron. 5. Sulphate of zinc. 6. Water acidified with a minute portion of hydrochloric acid. 7. Water poured on powdered arsenic. All these cylinders were connected with the positive pole, and were electrically united together by arcs of sheet copper, so that the same electrical current passed through the whole of them.

"After many months' action and consequent formation of certain crystalline matters, I observed similar excrescences with those before described at the edge of the fluid in every one of the cylinders except the two which contained the carbonate of potassa and the metallic arsenic; and in due time the whitish appearances were developed into insects. In my first experiments I had made use of flannel, wood, and a volcanic stone. In the last, none of these substances were present. I never for a moment entertained the idea that the electric fluid had animated the remains of insects or fossil eggs, previously existing in the stone or silica. I have formed no visionary theory that I would travel out of my way to support.

"In some cases these insects appear two inches *under* the electrified fluid, but after emerging from it they were destroyed if thrown back."

The insects also made their appearance in silicate of potassa four inches below the surface of the fluid, also in *fluo-silicic acid* two inches below the fluid. These experiments were repeated and others instituted with a similar view by the late Mr. Weeks of Sandwich. He passed currents of electricity through vessels filled with solutions of silicate of potash, under glass receivers inverted over mercury,—the greatest precautions having been taken to shut out extraneous matter; and in some cases the receivers were previously filled with oxygen gas. After an uninterrupted action of about a year and a half insects invariably made their appearance, exactly resembling those that occurred in Mr. Crosse's experiments some years previously.

In some of Mr. Weeks' experiments, the acarus made its appearance in solution of ferrocyanuret of potassium.

That gentleman, in a communication to the Electrical Society, stated that he had repeated these arrangements *without* electricity, placing the apparatus in every variety of position favourable to the development of insect life, but none appeared. Mr. Weeks considered these negative experiments as very important.

Another and later experiment made by Mr. Crosse deserves notice. He says:—

"I calcined black gun-flints in a crucible, and flung them while hot into water: I then dried and reduced them to powder. Of this powder I mixed one ounce, and intimately mixed it with three ounces of carbonate of potassa. I fused them together for five hours, increasing the heat, until it exceeded that necessary to melt cast iron. I removed the crucible, and then allowed the contents to become solid, which formed into a pale green glass. While still *hot*, I broke them into pieces: these *hot* pieces I threw into a vessel of boiling distilled water. I had previously prepared an apparatus to act electrically upon this fluid. It consisted of a common tubulated glass retort. The beak of the retort rested in a cup of pure mercury, from which proceeded a platinum wire, which passed up through the whole length of the retort, and when it reached the bulb was bent at right angles, so as nearly to touch the bottom of the bulb. The glass tube, which fitted air-tight into the neck of this retort, had a platinum wire passed straight through it, the upper part of which was hermetically sealed into the upper part of the tube, and the lower part of the wire was continued downwards. The two platinum wires were at a distance of about two inches from each other. When all was ready I poured the solution still *hot* into the bulb of the retort, thus affording a conducting medium between the two platinum wires, connected with the opposite poles of a small voltaic battery. An electric action commenced; oxygen and hydrogen gases were liberated; the volume of atmospheric air was soon expelled. Every care had been taken to avoid atmospheric contact and admittance of extraneous matter, and the retort itself had been previously washed with hot alcohol. This apparatus was placed in a dark cellar. I discovered no sign of incipient animal formation until on the 140th day, when I plainly distinguished *one* acarus actively crawling about *within* the bulb of the retort. I found that I had made a great error in this experiment; and I believe it was in consequence of this error that I not only lost sight of the single insect, but never saw any others in this apparatus. I had omitted to insert within the bulb of the retort a *resting place* for these acari (they are always destroyed if they fall back into the fluid from which they have emerged). It is strange that, in a solution *eminently caustic* and under an atmosphere of *oxihydrogen gas*, one single acarus should have made its appearance."

These insects also appeared in an atmosphere strongly impregnated with chlorine; but in this latter case they assumed the form of perfect insects, and remained undecomposed and unaltered for more than two years; in fact till the apparatus was taken apart; but, singularly enough, they never moved nor evinced any signs of vitality.

THE END.

9 781108 014915

Printed in the United States
By Bookmasters